Spatial Imaginaries in Mid-Tang China

Spatial Imaginaries in Mid-Tang China

Geography, Cartography, and Literature

Ao Wang

Cambria Sinophone World Series
General Editor: Victor H. Mair

CAMBRIA
PRESS

Amherst, New York

Requests for permission should be directed to permissions@cambriapress.com, or mailed to:
Cambria Press
100 Corporate Parkway, Suite 128
Amherst, New York 14226, USA

Front cover image: The twelfth-century "Map of Chinese and Foreign Lands," stone rubbing, 1903 (now in the Library of Congress, USA).

Library of Congress Cataloging-in-Publication Data

Names: Wang, Ao, author.

Title: Spatial imaginaries in mid-Tang China : geography, cartography, and literature / Ao Wang.

Description: Amherst, NY : Cambria Press, 2018. | Series: Cambria Sinophone world series | Includes bibliographical references and index.

Identifiers: LCCN 2018022098 | ISBN 9781604979411

Subjects: LCSH: Chinese literature--Tang dynasty, 618-907--History and critcism. | Geographical perception in literature. | Geography in literature. | Cartography--China--History.

Classification: LCC PL2291 .W36 2018 | DDC 895.109/003--dc23
LC record available at https://lccn.loc.gov/2018022098

Table of Contents

List of Figures

Acknowledgments

This book, which presents new research conducted after I completed my PhD program, would not have come to fruition without the help of my colleagues Wu Shengqing, Stephen Angle, Mary Alice Haddad, and William Johnston. The year I spent at the Newhouse Center for the Humanities at Wellesley College was crucial for the research and writing of the manuscript. My friends and colleagues Tian Xiaofei, Anna Shields, Qian Nanxiu, Zhang Ling, Linda Feng, Huai Rui, and Song Mingwei offered me numerous insightful comments and suggestions to improve this book. Eleanor Goodman, Allison Deventer, and Richard Stack copyedited the manuscript at different stages, and I am very grateful to them. Eleanor, in particular, is also a fellow poet, co-translator, and my closest friend in the United States.

I want to thank my mentors and teachers Robert Hegel, Kang-I Sun Chang, Michelle Yeh, David Der-wei Wang, and Wang Yugen for their ongoing guidance since my twenties. I am also grateful to Toni Tan and David Armstrong of Cambria Press for shepherding this book to publication with their exemplary work. This book is dedicated to Miya Qiong Xie, who reads each word I write, gives me hope, and supports my every dream.

Spatial Imaginaries in Mid-Tang China

Introduction

Interactions Between Geographical Advancements and Spatial Imaginaries

This book studies the interactions between geographical advancements and spatial imaginaries in literature during the mid-Tang period (roughly from the 790s to the 820s), a topic that has never before been examined in detail. I argue that during the period of imperial reconstruction following the chaos of the An Lushan Rebellion (755–763), geographical advancements and literary imagination joined forces to capture a radically changing world and to give that world new meaning. By geographical advancements, I mean progressions in the geographical understanding of space, which include but are not limited to the accumulation of new geographical knowledge. I use the term as a broader concept that also encompasses a heightened geographical awareness, new spatial perspectives on the world, and new ways of thinking about human inhabitation. These advancements were acquired via surveys and explorations out in the field, and they were embodied in maps and geographical works of varying scope that reflected these surveys and explorations.

The conceptual counterpart of this phenomenon within literature I call "spatial imaginaries." I use this term to describe the imaginations and representations of inhabited and uninhabited spaces on different scales. Through a thorough consideration of geography and literature in the mid-Tang period, and their intersections and juxtapositions, this book aims to capture the affinity and interplay between these two important intellectual pursuits at a crucial transitional period in premodern China.

Affinity and interplay are key terms that I use to define the sophisticated relationship between these two fields of spatial representation. We often understand the relationship between different disciplines as one of influence. The rise of one discipline can spark methodological changes in another, transforming the ways a group of specialists think about their own work. The interplay I am about to present between the literary and geographic fields in the medieval era is different. This difference comes first and foremost from the fact that the masters in these two fields in mid-Tang China were essentially one and the same group of multitalented people. As they pushed the frontiers of world exploration ever further, they created—consciously or unconsciously—a new contextual space in which texts in both fields were produced and consumed. From this arose an affinity between literature and geography: they may have shared similar perspectives on space, intertextualized with each other in the process of composition, or drawn inspiration from the same human feats in a broadening geographic world. This is not a narrative of a direct influence of one discipline on another, a relationship that the surviving sources at this point do not allow us to reconstruct. Nevertheless, the recognition of a cross-field affinity allows us to productively examine certain literary texts of this particular era in the light of new spatial imaginations, perspectives, and conceptions.

A metaphor about image formation from physics may help better illustrate this relationship. A projected image is created by a beam of light that emanates from an original object, travels through an optic lens, and then hits a receiving surface. The light, transmitted through and refracted by the lens interface, results in a changed image of the original on the final surface. Let me appropriate this notion from physics to explain my idea of cross-field affinity. The final surface on which the projected image appears corresponds to the field of literature. The images projected onto it are the spatial imaginaries, which may take various forms, from the pictorial and graphic to the textual, mental, and so on, or any combination thereof. The lens interface that transmits and refracts the light beam corresponds roughly to an ontological and epistemological filter through which writers processed information to make sense of the world. In premodern China, this refracting lens is essentially cross-field, although at different times and for different minds, the composition of that lens varied widely.

The focus of this study is how space is processed through this intellectual lens and transformed into images on a literary surface. My approach is to investigate how the literati's geographical awareness and literary sensibility, two of the many dimensions that constitute this filtering lens, worked together to produce some of the most sophisticated spatial imaginaries in Chinese literary history. The prominence of geography in this study is a reflection of the significance the field had during the mid-Tang, when efforts to rejuvenate the post-Rebellion Tang Empire hinged in large part upon a better geographic understanding of its changed frontiers and underdeveloped hinterlands. The intellectual filtering lens, dually sharpened by geography and literature, enabled writers to delineate a rapidly changing imperial space and beyond with a mix of grandeur and subtlety, enchantment and accuracy. Their spatial imaginaries reflected the ambition and anxiety of contemporary men of letters, who aspired to be pillars of the imperial enterprise.

Given that the past millennium has destroyed or obscured most geographical materials from the Tang, we can reconstruct the geographical dimension of this filter only in indirect ways. In fact, here I have had to rely heavily on different kinds of textual records of geographical materials, many rich in literary elements, to restore the facts, context, and effects of the mid-Tang geographical advancements. In this sense, I treat literature as both the bearer of geographical imaginations and the carrier of traces of lost geographical works. That said, literature was by no means merely a passive reflector of a geographically filtered world. Mediated also by distinct literary sensibilities, the spatial imaginaries had a creative edge that further animated the elements of geography living on in literature. This creativity transformed measurement and knowledge into expressions of human emotions and imagination.

Moreover, my observations not only concern the field of geography and literature but relate to other media as well. In my study, the receiving side of the projected images is literature. However, the same spatial imaginaries were also projected onto different media and were mediated by the same ontological and epistemological filter. Accordingly, the spatial imaginaries in literature that were inspired by a cartographical eye or a geographical awareness also interacted with and spoke to a spatial consciousness being addressed in multiple representational modes. As a result, in the geography-informed spatial imaginaries in literature, we not only observe an interplay between geography and literature, but are in fact initiated into a new worldview that emerges on a cross-media intellectual horizon. In this sense, what I find in my literary study may also apply to products of other representational modes of the mid-Tang, such as painting, steles, religious artifacts, and so on. I will therefore extend my discussion to media beyond literature when relevant.

This topic first drew my attention when I found curious overlaps between geography and literature in the mid-Tang period. Literary masters such as Han Yu 韓愈 (768–824), Liu Zongyuan 柳宗元 (773–819), Bai Juyi 白居易 (772–846), Yuan Zhen 元稹 (779–831), and Liu Yuxi 劉禹錫 (772–842) all had a close relationship with contemporary geographic developments. Yuan Zhen was a distinguished mapmaker and submitted several maps to Emperor Muzong 穆宗 (r. 820–824). Han Yu served as the Vice Director of the Bureau of Operations (*zhifang yuanwailang* 職方員外郎) in the Ministry of War (*bingbu* 兵部), where his duties included managing imperial maps. Liu Zongyuan, Bai Juyi, and Liu Yuxi all repeatedly evoked contemporary maps and geographic guides in their writing about landscape, history, and specific local sites. Although literature and geography were never mutually exclusive disciplines in premodern China, the acute collective awareness of the importance of geography among the mid-Tang literary masters is still a curious departure from earlier literary norms. Their frequent and explicit invocation of geographic terms and works in their literary writing gives further proof of the vitality of the mid-Tang confluence between the two fields.

As more evidence and clues surfaced in the course of my research, I became convinced that certain significant achievements in geography and literature in the mid-Tang were closely related to each other. The period from the An Lushan Rebellion to the mid-Tang is one of the most crucial eras of geographical reconstitution in Chinese history. Wars and economic decay in the northern heartland of the empire drove massive populations to unfamiliar peripheries and compelled the educated elite to seek more geographic knowledge and tools for navigation. Strong political and intellectual calls to revitalize the much weakened, shrinking post-Rebellion empire added further symbolic value to the study of geography. Many leading political and cultural figures considered the expanding scope and depth of geographic knowledge to be a positive sign pointing to a successful imperial revival. This heightened geographic awareness across the empire is illustrated by the fact that the two long-standing chief ministers at the turn of the ninth century, Jia Dan 贾耽 (730–805) and Li Jifu 李吉甫 (758–814), were both acclaimed geographers. They produced a dazzling array of works such as maps and imperial geographic encyclopedias that became milestones in premodern Chinese geography, and their emphasis on geography became a shared basic tenet among mid-Tang cultural elites.

Correspondingly, mid-Tang literature witnessed innovative ways of exploring and representing space, from the vast cosmos to an obscure cave by an unknown river. The poet Li He 李賀 (790–816), for example, depicted a bird's-eye view of the entire empire from the vantage point of the cosmos in his poem "Dreaming Heaven" ("Mengtian" 夢天), whose wild and innovative imagination has impressed many critical readers. Detailed and delicate representations of the localities within the empire also flourished in the mid-Tang period. A wide range of poems in various forms by Liu Yuxi, Bai Juyi, and Yuan Zhen, among others, are known for their artful depictions of local geographies, while genres such as the office inscription (*tingbi ji* 廳壁記), or prose records written by officials or their friends about local territories and inscribed on office walls, became popular at the time. Even more prominent was the landscape essay (*youji* 遊記), a literary genre that first took its full form in the mid-Tang period and had a profound influence on the prose writing of later times. This genre was marked by its aesthetically pleasing yet geographically precise representations of the landscapes in the southern regions of the empire. These diverse literary forms all opened up new spatial imaginaries in Chinese literary history.

What bridged the extraordinary achievements in the fields of geography and literature in the mid-Tang period was a group of multitalented literati, whose deep involvement in contemporary geographical advancements allowed them to apply a geographical filter seamlessly to the making of literary imageries, metaphors, and narratives. They internalized cartographic images made available through contemporary map-making and map-reading activities to create innovative views of the world, or to navigate swiftly through vast spaces in literary texts. They intertextualized and interacted with local geographical works, most prominently map-guides (*tujing* 圖經), when writing about the diverse and intricate local spheres of the empire. When they were driven to the empire's uncharted frontiers due to wars, mass migration, political demotion and other social changes, they made the exploration, domestication and inhabitation of the newly developed land and landscape a literary topic, and thus invented brand-new literary genres and styles. As such, these mid-Tang writers, themselves enthusiasts of geography, channeled contemporary geographical advancements as they engaged with space through literature.

At the same time, the mid-Tang writers' powerful reworking of geography brought new meanings to spaces, places and sites, and offered new materials for future geographical writing. In literature, authors reorganized natural objects, local sites, and other knowledge recorded in geographical works into images or narratives that were conducive to aesthetic, political, or social values. For them, geographical information meant not only knowledge and perspectives, but also words and letters to be matched with rhymes and woven in parallelisms, or rhetorical devices that would empower the authorial voice in their literature. For these purposes, they would sometimes depart from the verifiability of geography, giving it a fictional, creative twist; at other times, they established connections between otherwise unrelated places, thus subtly altering readers' spatial perception in related regions. Furthermore, by writing about landscape and inscribing their writings in the landscape itself, literary authors also helped establish new landmarks in the physical space and changed the milieu of the local environment. These landmarks would then in turn feature in later geographical works and become famous sites of literary pilgrimage.

To summarize, by identifying traces of the epistemological and aesthetic filter in part shaped by contemporary geographical advancements in mid-Tang literary texts, my cross-field investigation will probe into the sources of inspiration for those texts' diverse and intricate spatial imaginaries and illuminate their effects on readers. This geographical perspective will give birth to a new and revealing line of interpretation into some of the most important works of Tang literature. More broadly, a close investigation of how geography and literature cross-pollinated in the mid-Tang period will uncover a rich cultural history of the intellectual exploration of the world on both fronts. In a time when the empire was struggling to revive itself, multitalented literati used their geographically informed literary writing to address their ambitions and anxieties, and to cope with the rapidly changing world.

SPATIAL IMAGINARY

The main object of analysis in this book is the spatial imaginary in mid-Tang literature. An imaginary is the process of image-making. The notion of an image, in my book, refers to the mental schema or representation of the real through a wide range of media. The image as such can be graphic, pictorial, verbal or mental; it is in most cases visual but can also be multisensory. In W.J.T. Mitchell's seminal work on the image, he conceives the various kinds of images as a family tree.[1] Mitchell notes that people tend to think of graphic and pictorial members of the family as images in the strict, literal, or proper sense, while considering verbal and mental images to be their illegitimate, metaphorical offspring. However, for Mitchell, all members in the family are equally images and share much in common. Graphic/pictorial images are not easily distinguished from other kinds of images, as "they are not stable, static, or permanent in any metaphysical sense; they are not perceived in the same way by viewers any more than are dream images; and they are not exclusively visual in any important way but involve multisensory apprehension and interpretation."[2] Mitchell's insight allows us to put different kinds of images on the same horizon for analysis.

This broad definition of image is particularly useful for my study of the spatial imaginary in medieval Chinese geography and literature. Mitchell's schema removes the hierarchical division between pictorial/graphic image and verbal/textual image. One key feature of the set of geographic and literary work studied here is that pictures and texts work collaboratively, and sometimes interchangeably, to represent the physical world. In his seminal study of Chinese cartography, Cordell Yee notices that "the distinction between word and visual image, so strong in the Western tradition, is not nearly as sharp in China....The usual oppositions between visual and verbal, cartographic and pictorial, mimetic and symbolic representation may not apply."[3] Yee then attributes this characteristic of Chinese cartography to the fact that "the Chinese intellectuals who made and read maps held 'broad learning' (*boxue* 博學), not specialization, as an educational ideal and seem to have regarded cartography in those terms."[4] What Yee observes for Chinese cartography also applies to traditional Chinese geographic studies in general. It is commonly known that medieval geographic works often included a pictorial element as well as a textual element, and the maps or pictures themselves were heavily annotated with text. Moreover, when medieval Chinese geographers drew maps or wrote geographic works, in addition to observational measurements of physical geographical objects, they also relied on texts as important sources of information.[5] In these ways, visual and textual materials supplemented and substantiated each other in conjuring up comprehensive geographical images. Consequently, in literary texts that share similar ways of seeing the world as geographic works or that draw reference from them, one sometimes observes a "verbal map": the text lays out the spatial arrangement of geographic objects and navigates the reader through the objects as a picture or a map would do. In a broader sense, a verbal map may be found in any literary work dedicated to delineating a physical space. But as our discussion in the book will show, works by medieval Chinese authors often bear a resemblance to, or can be easily associated with, the textual parts of Chinese cartographical works, thus aligning them even more closely with actual maps.

Moreover, Mitchell's emphasis on mental images as a proper member of the image family applies to medieval Chinese landscape literature. Landscape literature, especially the landscape essay, was a representative literary genre that engaged with space during the mid-Tang period. As Xiaofei Tian points out, early Chinese medieval literature and art depict the landscape as *xiang* 象, which refers to a kind of "mental seeing and image-making."[6] For early medieval literati, "landscape was essentially a grand image (*xiang* 象), and the perception, interpretation, and indeed the very construction of this image were contingent upon the workings of the mind."[7] Accordingly, the rise of the representation of the natural world in literature and painting at the time can properly be seen as an "extension of the primary engagement with the inner world of an individual."[8] According to Tian, this way of seeing the world with the mind's eye established a textual and visual tradition in Chinese literature and painting that was carried out well into the modern era.[9] The mid-Tang literati and landscape writing that I will discuss here came under this influence, and this mind's eye is critical to my analysis of the mid-Tang literati's geographic and literary work. The eye sometimes takes on the form of the imposition of a conceptual image onto real natural terrains; at other times, it manifests itself as the author's retrospection as enabled via observation of a landscape. In any case, it was sometimes the author's own mind's eye that guided him through the discovery, construction, and writing of the landscape, all of which changed the understanding of local geography for these writers' contemporaries and later generations as well.

Image is, by definition, cross-media and cross-disciplinary. It can certainly be found in a landscape painting, or in a map that represents space graphically. These are perhaps forms of image that are closer to the "image proper" in Mitchell's schema. A spatial image can also be conjured up in literary or geographic texts through verbal descriptions; religion or philosophy, too, have their ways of making images in mental, verbal, or visual form. After all, boundaries between media and disciplines were not clearly demarcated in the premodern Chinese intellectual world. Geographic advancement, literary creativity, and religious contemplation may complement and reinforce one another within a single intellectual work. Therefore, in my book, I treat the spatial imaginary in mid-Tang literature as a phenomenon that is both multimedia and interdisciplinary in itself, and which is in constant interaction with other media and disciplines. In other words, I pay special attention to how the literary imaginary blends in elements from paintings, maps, geographic works, and religious texts on the one hand, while on the other hand always maintaining an awareness of its own relationship to image-making in other intellectual fields such as geography and religion.

Finally, as mentioned, images are not exclusively visual but involve multisensory apprehension. The exploration of the multisensual dimensions of geographic objects is an important component of the mid-Tang literary works under discussion here. Poetry that addresses a map-reading experience may evoke a wine cup, which indicates that the comprehension of a map image could be an organic part of a social gathering, together with a banquet and the drinking of wine. Landscape essays that record the creation of new landmarks in the imperial south delineate not only the landscape in its visual aspect, but also the changing spectrums of sound, temperature, and other sensory geographies of a given site. In Yuan Zhen and Bai Juyi's poems about their exilic experiences, the malarial air, the noise of swarming insects, and the clamorous local people represent the imperial peripheries as simultaneously dangerous and exotic. This all creates a verbal spatial imaginary that requires the reader to not only visualize the scene, but also conjure it up through his/her full sensory associations. Only then can the reader fully appreciate the spatial dimension constructed in the literary work.

Having defined the term "imaginary," let us now move to the term "spatial." Relating to the spatial is a set of words and notions that need to be defined: space, place, landscape and site. In my book, space serves as an inclusive concept. As such, it can be real or imagined, material or textual, literal or literary; it can entertain different scales, from the cosmos to the wall of an office. Yet it is always socially constructed and is imbued with human meanings and values. My definition is different from the conventional distinction between space and place in humanistic geography and is closer to the poststructuralist understanding of space as social construction. In humanistic geography, space is conventionally considered to be something more abstract and less defined than place. When a space is marked off and is given meaning, value, or identity, space is transformed into place.[10] As Yi-fu Tuan puts it, "Compared to space, place is a calm center of established values."[11] In recent decades, however, this differentiation has been sometimes confused by the idea of the social space, which stresses that space as perceived by humans is socially constructed and is, like place, imbued with powers and values.[12] This idea expands the denotation of the concept of space and makes it difficult to distinguish space and place according to whether or not human meanings and values are given. For literary studies, the expanded meaning of space makes more sense, since the moment it is represented in literature, it is culturally defined and is colored by human experience. This is even more the case when we speak of the space that literature not only represents but also recreates, for example fictional space or literary space. In the present study, space and place of all kinds are heavily loaded with social, cultural, or political meanings.

Nevertheless, it is helpful to maintain a certain distinction between space and place. For my purposes, I find Yi-fu Tuan's rather figural description more useful: "If we think of space as that which allows movement, then place is pause; each pause in movement makes it possible for location to be transformed into place."[13] Or, to put it slightly differently, if we imagine space to be the overall range of a map, then places are dots on the map. A place can in turn have its own map, and dots on that map, and so the distinction between space and place is always relative and transitory. In my literary study here, these dots stand for the human desire to demarcate and mark a location, and to fix it against forces of ongoing movement. Understanding this desire is indispensable to our interpretation of the spatial imageries in literature. These place-making activities sometimes take place in a real space and are then represented through literature; equally, literature may function as the decisive force to mark a location in a real space and consequently transform it into a place. In any case, for any of the forms of space I discuss, place is its relational concept that signifies a mark, a pause, a node, which also makes us further aware of the openness and movement of space.

Following the concepts of space and place, I introduce two more specific spatial notions: landscape and site. Landscape is the world as it appears to the human eye. Tim Cresswell summarizes landscape as "a portion of the earth's surface that can be viewed from one spot."[14] In medieval Chinese geography and literature, landscape corresponds to the term *shanshui* 山水, which literally means mountains and rivers, two of the most basic elements of nature. In other words, landscape in this context primarily refers to natural landscape; the buildings included in the landscape are mostly pavilions, terraces, and other constructions that are designed to facilitate human appreciation of the landscape. In addition, as I have mentioned earlier, the landscape in the premodern Chinese intellectual world was not only to be viewed, but also to be heard, smelled, felt, and sometimes inhabited. When literary authors engaged with landscape in their works, the landscape was almost always associated with some form of introspection, such as a philosophical exploration of the relationship between humans and nature, or an intellectual dialogue with the minds of earlier landscape writers.[15]

The term "site," on the other hand, emphasizes human action. In the Merriam-Webster Dictionary, site is defined as "the spatial location of an actual or planned structure or set of structures (as a building, town, or monuments); a space of ground occupied or to be occupied by a building; the place, scene, or point of an occurrence or event."[16] In short, site is the location or space for buildings and events. Landscape and site are not mutually exclusive concepts, but rather closely overlap. A site can be part of a landscape, and vice versa. In my book, a site can be a shrine where religious activities are carried out, or a pavilion where literati liked to hold social gatherings with a component of landscape viewing, or a scenic spot marked out in a literary work. When a site functions to mark a place of human action in a broader space, that site becomes a landmark. Literary engagements with sites are just as multifarious: a literary work can record human activities at the sites; the inscription of a literary text in a landscape can create a new site; and in cases such as orisons of appeal (*zhuwen* 祝文), refined literary texts are read aloud as part of the rituals associated with religious sites. In all of these engagements, the focus is always human activities in the space.

In Chinese intellectual traditions, different types and scales of space, place, landscape, and site invited different geographic and literary engagements, and were associated with different social, cultural, and political meanings. In the following chapters, I will delineate the contours of the geography-informed literary semiotics of the various spaces in the mid-Tang era.

THE ENCOUNTER BETWEEN GEOGRAPHY AND LITERATURE

In terms of its primary intellectual arguments, my book deals with the relationship between geography and literature, and the meaning of this relationship for both sides of that dynamic. Scholarly interest in the interaction between literature and geography has long existed. In our contemporary disciplinary field, this topic can be most easily located under the category of literary geography. Ever since the modern field of literary geography was initiated in the 1970s, there have been abundant academic discussions regarding how the two disciplines can benefit each other. In the following, I would like to engage with these discussions both in the West and in China, so as to illuminate the main methodology of my work, as well as its contribution to broader scholarly trends.

In the 1970s, humanistic geography started to attract attention from geographers as an approach that puts human experience and human subjectivity firmly at the center of geographic studies. In seeking new materials from fields within the humanities, some humanistic geographers found that literature is often "a valuable source for examining more subjectively the sense of place and could provide accounts of personal appreciation and experience of landscape."[17] While geographers hailed the importance of literary sources in this new field, however, literary scholars remained conspicuously absent from the supposedly joint enterprise for decades. Geographers accomplished much in these years of lonely adventure, but the geography-centered approach has also been criticized for its instrumentalization of literary texts. As Marc Brosseau points out, geographers have mostly used literature to support or provide answers for their particular geographical quests and have spent little time on the literary texts themselves.[18] As a result, they have limited themselves largely to realistic depictions of places and environments while losing sight of literature's potential to offer other types of geography through its inherent form and its singular use of language. Scholars have therefore called for a more dialogical relationship between geography and literature, one that pays due attention to the poetic language and form of literature.[19]

In recent years, we have witnessed increasing scholarly efforts to make the interaction between literature and geography a two-way street, most prominently to consider geographical and literary texts in light of each other. For example, the geographer Edmunds Bunkse, in his book *Geography and the Art of Life*, combines geography and literature to recall his World War II childhood experience. He not only evokes literary works to help him articulate his subjective experience of different places and landscapes, but also demonstrates that scholarly geographical work can be literature, and therefore may be subjected to the same critical reading process that literary scholars have used for poetry and novels.[20] Literary scholars, on the other hand, have analyzed the fictional and narrativized "spaces" in literary texts through a juxtaposition of literary descriptions with real geographic spaces, invoking principles of geographic spatial analysis in the process.[21] Building upon these efforts, Sheila Hones, the author of several recent monographs on literary geography, suggests a thorough integration of the two disciplines. She proposes that an ideal work of literary geography "has to be done in such a way that neither the literature (the texts and the study of those texts) nor the geography (the world and the study of that world) become reduced to the status of subject matter, theme, or raw data. In order to achieve this interdisciplinary balance, literary geography has to go beyond the literary analysis of geographical themes or the geographical analysis of literary texts. Literature and geography have to function as a combined double subject, on the one hand, and a combined, double, theoretical, and methodological framework, on the other."[22] In other words, we may say that both literature and geography need to be treated as text and context at the same time. They are both central objects of investigation; simultaneously, the investigation in one field may be conditioned by theories, perspectives, and knowledge from the other.

In China, the term "literary geography" (*wenxue dili* 文學地理) was first introduced in 1902 by the modern scholar Liang Qichao 梁啟超 (1873–1929). In an article titled "A Discussion of the Main Trends in Chinese Geography" ("Zhongguo dili dashi lun" 中國地理大勢論), Liang discusses how literature, art, and scholarship in the northern and southern areas of China took different shape in part due to the different physical and cultural environments in which they were rooted.[23] Liang's approach to regional literature and culture does not fall neatly into the basket of interdisciplinary literary geography that we have outlined earlier. He used geography as an explanatory factor for the theme and style of various literary works, but he cared little about how literature could improve our understanding of the geographical environment or the literati's spatial consciousness in the historical periods under discussion. Nevertheless, it does show a leading twentieth-century Chinese intellectual's awareness of the geographical dimension of China's cultural and literary history. In terms of the disciplinary history, Liang's work defined the fundamentals of Chinese literary geography for the following decades. Even today, despite a radical expansion of the field in China, the basic emphasis on the geographic distribution of literary elements across regions remains largely unchanged.

Zou Jianjun 鄒建軍, an important contemporary scholar of Chinese literary geography, has proposed a list of topics that he believes the field ought to include, ranging from the influence of the natural environment on literary writers and phenomena in literary history, to spatial imaginaries in literary works, to the influence of geographic discovery and outer-space exploration on literary texts.[24] But his theoretical proposal has yet to find materialization in the practices of academic research. For example, in his book-length study *The Study of Literary Geography* (*Wenxue dilixue yanjiu* 文學地理學研究), Zeng Daxing 曾大興 defines the scope of his study as "the geographic distribution, combination, and evolution of literary elements; the regional characteristics and differences of literary elements and the overall shape of literature; the relationship between literature and geography." His investigation includes how literary writers, schools, activities, styles, and other phenomena are distributed across the different geographic regions in China and are influenced by the regional milieu, and how they have in turn had an impact on regional cultures.[25] *A Comprehensive Study of Literary Geography* (*Wenxue dilixue huitong* 文學地理學會通) by Yang Yi 楊義 further adds the geographic routes of the dissemination of literary works and influence to the discussion, but the region still works as the basic unit on his map of literary dissemination.[26] The regional approach in the study of literary geography primarily takes geography as context and method. It places the focus on the regional geographic environment surrounding literature and adopts analytical tools typically employed in the field of geography, such as phenology, urban design, and so on.

Some other studies, although not framing themselves as literary geography, explore the intertextuality between geographical and literary texts. Among them, two deal with the Tang period. The book by Liao Yifang 廖宜方 on the memory of history in the Tang Dynasty discusses how Tang literati engaged with pre-Tang cultural history in their geographical and literary writing, and how they intertextualized with each other's works.[27] Wang Liqun 王立群, in contrast, studies how geographical writing in the early medieval period informed the emergence of the literary genre of the landscape essay in the mid-Tang period.[28] These two works suggest the fluid relationship between geography and Tang literature on a textual level, thus demonstrating refreshing approaches to the study of the interdisciplinary encounter between geography and literature in medieval China. Nevertheless, for a book on cultural history, Liao's work does little close readings of literary texts in terms of their language and style; while Wang's book is primarily focused on the literary incorporation of geographical texts produced at a much earlier period, not those from the time.

Drawing inspiration from this large body of Western and Chinese scholarship, my work departs from conventional studies of Chinese literary geography and echoes Sheila Hones' proposal for an integrated approach to literature and geography. It explores new frontiers in the field from a medieval Chinese perspective by capturing a specific historical moment when fields and knowledge were not yet demarcated by discipline, and when experts in geography and literature were one and the same group of people. First of all, my subject requires a broader understanding of geography and space. As I briefly touched upon at the beginning of this introduction, I understand geography not as a convenient synonym for geographic environments, nor merely as a body of knowledge in and about such environments. Rather, I employ a richer conception of geography that includes the literati's heightened awareness of the urgent need to know more about their lived space, their direct involvement in exploring and recording that space, new perspectives in seeing and representing the world that issued from their geographical activities, and new ways of thinking about human inhabitation. Furthermore, geography in my study is not only spatial. It also has a temporal dimension. As my later discussion will show, premodern Chinese geographical studies are known for their wide-ranging coverage of local histories and customs. Accordingly, I am interested in how geography as a medieval idea subsumed history under new, massive imaginaries of space-time, shaping the experience of the viewer/visitor at any given time. I believe that this holistic perspective should enable a fuller appreciation of the complex encounter between geography and literature in the Chinese medieval world.

The relationship of geography to literature also goes beyond a unidirectional influence. In my study, I recognize the two fields, to borrow Hones' terms, as truly a "combined double subject, on the one hand, and a combined, double, theoretical, and methodological framework, on the other."[29] They are a dual dimension of the multitalented literati's intellectual outlook. In my work, the combined and double framework matches nicely with the mid-Tang literati's new cultural-spatial outlook, which emerged as a result of their exploration of the world on both the geographical and literary fronts following a major geographic-political transformation of the empire. This cross-field framework is important because it constituted a vital part of the intellectual context in which these literary works were written and read. I investigate spatial imaginaries in literature within this framework in order to explore new meanings, interpretations, and implications of those imaginaries. In addition, many geographical works of the time are no longer available in their original form and can only be gleaned through records or fragments preserved in literary texts. As a result, literature, especially that which concerns space, also offers crucial referential materials through which the contours of the cross-field exchange can be at least partially reconstructed.

Specifically, the cross-field affinity I examine may take various forms in the literary texts under examination. Contemporary geographical advancements often enter literature in the form of awareness, experience, perspective, or knowledge. Sometimes a direct influence is identifiable, for example when authors of literary texts evoke geographical works or terms explicitly to enhance the rhetorical or poetic power of their writing. At other times, contemporary literary and geographical texts intertextualize with each other by sharing similar thematic and stylistic features. When a cartographical perspective, rather than specific knowledge, is involved in literature, the literary texts are often found to exhibit ways of structuring imageries or configuring metaphors that resemble those underlying contemporary maps. In still other cases, literature itself records contemporary geographical explorations on both the collective and individual levels. A thorough examination of these diverse, fluid cross-field relations restores the geographical dimension of the overall intellectual outlook of the time to a greater extent than the surviving geographical materials alone can do. It also gives due attention to both the aesthetic and epistemological significance of the formal elements of literature.

In today's world, where the production of knowledge is highly specialized and professionalized, the Chinese medieval case that I study here suggests an alternative and provides new inspiration for interdisciplinary exchange to scholars of our time. There has never before been a systematic study of the intellectual encounter between these two fields in premodern China from the perspective described earlier. My work will fill this scholarly lacuna and enrich studies of literary geography in China and beyond.

ORGANIZATION OF THE BOOK

This book is divided into five chapters. The first chapter provides a necessary historical overview that will set the stage for the discussions that follow. I start with a brief description of the interplay between geography and literature in Chinese history predating the mid-Tang. I then introduce the geographic developments of the Tang in detail, laying out the protagonists, technologies, methods, and theories involved. In particular, I provide thorough background information on the three geographical works of the mid-Tang that are directly involved in my later discussion of literary texts: the grand map "Map of Chinese and Foreign Lands within the Seas" ("Hainei huayi tu" 海內華夷圖, often referred to as the "Map of Chinese and Foreign Lands" ["Huayi tu" 華夷圖]) by Jia Dan, and the *Maps and Treaties of the Provinces and Counties of the Yuanhe Reign* (*Yuanhe junxian tuzhi* 元和郡縣圖志) by Li Jifu.

Chapter 2 studies the creation of some spectacular images of the world in mid-Tang literature through an intermedial investigation of literature and maps, in particular Jia Dan's "Map of Chinese and Foreign Lands." Literary representations of the world have been taken as a subject in the well-ploughed field of literary history. This chapter introduces a cartographic approach to these representations, which have been somewhat neglected in their own right and have received even less attention in relation to literature. In the chapter, I describe the interaction between maps and literature in these cases as a conversation. On the one hand, the imperial grand map representing the entire known world afforded writers fresh perspectives with which they could imagine the cosmos and the empire in exciting new ways. On the other hand, the literary texts also speak back to these maps, highlighting the social, cultural, and aesthetic features that are integral dimensions of Jia's map but could not be expressed directly, explicitly, or fully on the map itself.

My analysis begins with a brief discussion of the high-Tang poet Du Fu's poem "At a Banquet Lord Yan Held in His Hall, We All Composed Poems on a Painting of the Cliff-Paths of Sichuan, and I Was Assigned the Rhyme Word 'Kong'" ("Yangong tingyan tongyong shudao huatu de kongzi" 嚴公廳宴同詠蜀道畫圖得空字). Although the poem was produced before the mid-Tang period and the map was merely a regional map, in this poem I will identify some general patterns of map-reading poetry that will reoccur in its mid-Tang successors. I then move on to analyze works by three mid-Tang writers, Li He, Liu Zongyuan and the slightly later poet Zhang Hu 張祜 (792–853), in relation to the advancements in cartographical skills in the same era. I will analyze Li He's poem "Dream Heaven" in light of some cartographical effects brought to the fore by Jia Dan's map, such as its cosmic scope and historical depth, and propose to read the poem as one that mimics a map-reading experience. I will then discuss several of Liu Zongyuan's poems and one of his landscape essays, "Record of an Excursion to Huang Creek" ("You Huangxi ji" 遊黃溪記), all written during his demotion in the south. I argue that Liu's creative engagement with cartographical objects, terms, and perspectives not only brings the spatial imageries in his writing added delicacy and precision, but also facilitates his political negotiation with the central court from a peripheral perspective. Finally, I will turn to two poems about "Maps of Mountains and Seas" by Zhang Hu. I explore how in his work the poet imposed a grand map view onto the ancient map-pictures and argue that the painterly nature of Jia's grand map was key to Zhang's poetic innovation. All these examples signify the magnitude of the literati's large-scale spatial imagination at the time and imply a geographically informed observer of the world behind the words.

While chapter 2 deals with grand images of broader spaces, chapter 3 narrows the focus to enter into the complex depths of the empire. This chapter examines how the literati's familiarity with regional geographical references, most prominently the map-guides, informed their creative usage of local knowledge in various genres and locations of cultural production and performance. The mid-Tang period witnessed a new literary trend that featured the employment of diverse and detailed local knowledge in literature. Examples range from long travel poems and fashionable prose genres such as orisons of appeal and office inscriptions to poetry, prose records, and fantastic stories. All these forms of literary writing engaging with the local sphere point to a common epistemological basis of local geographical knowledge, which I argue is the map-guide. Accordingly, in these literary texts, we observe a process of cross-field multilayering of textual knowledge between geography and literature.

The case studies offered in chapter 3 will start with Yan Zhenqing's 顏真卿 (709–784) two records about local Daoist sites written when he was in service in Fuzhou (in present-day Jiangxi Province) and then inscribed in stone. Like the example of Du Fu in chapter 2, Yan Zhenqing's work also preceded the mid-Tang but anticipated later local writing in terms of how they evoked map-guides to take account of local spiritual traditions and to strengthen the author's political argument as embedded in his writing. Following Yan's work will be a discussion of Liu Zongyuan's office inscription about Wugong County and of Liu Yuxi's long poem about Liyang, which he also later inscribed on the wall of his local office. In both cases, the two writers selectively rephrased the inclusive but fragmented geographical knowledge in the map-guides into coherent narratives of local geography that they would further utilize to bolster their political criticism of the central court. In addition to office inscription and other writings inscribed on the office walls, another large body of local writing that interacted with the map-guides consisted of orisons of appeal and other texts engaging with the governance of local deities and their shrines. I argue that all of these works rely heavily on the geographical information provided by the map-guides; at the same time, we sometimes find interesting stories in the map-guides that were likely intended to subtly manipulate the official's view of widely worshiped but illegitimate local deities. The chapter will end with examples of travel writing with explicit reference to the map-guide by Han Yu and Wang Jian 王建 (767–830), as evidence of yet another use of the map-guides among the mid-Tang literati.

Chapter 4 provides a new reading of some of the most important landscape essays about the southern part of the empire written by Yuan Jie 元結 (723–772), Liu Zongyuan, Han Yu, and Liu Yuxi. Because most of these landscape essayists wrote in the underdeveloped south to which they had been sent as political punishment, many critics believe that the harsh political dynamics of the period were the driving factor behind their favorable representations of the landscapes they encountered there. My reading supplements previous scholarship by situating landscape writing within the context of the mass southward migration after the Rebellion and the consequent geographical transformation in the south. It explores how literature helped establish new homes and create new landmarks in the developing imperial frontiers and served to transform the unfamiliar land into dwelling places. Many of these landmarks were then included in later geographical works and became an essential part of local geography. The chapter provides fresh interpretations of classical landscape essays from the perspective of home-building in the midst of radical geographical change and outlines a relationship of mutual reinforcement and inspiration between literature and geography.

I begin with a brief overview of the radical demographical and geographical changes that the imperial south underwent after the Rebellion. I then discuss works by Yuan Jie, a writer-official active shortly before the mid-Tang, as the prototype of the mid-Tang landscape essay. I argue that central to Yuan's landscape writing was an image of a home in the south that could meet a northerner's standard, one that combined personal, practical, and aesthetic needs with communal benefits. Following in Yuan Jie's footsteps, mid-Tang writers such as Han Yu, Liu Yuxi, Liu Zongyuan, and other lesser-known literati took up the task of transforming the desolate south into an inhabitable space by establishing new home-like landmarks, writing about them, and then inscribing their writings in the landscape itself. Finally, I will conduct a revisionist close reading of several of Liu Zongyuan's most acclaimed landscape essays, such as "To the Little Rock Pond West of the Little Hill" ("Zhi Xiaoqiu xi Xiaoshitan ji" 至小丘西小石潭記) and "Rocky Stream" ("Shijian ji" 石澗記) within this new context, interpreting these works beyond the personal scope of political frustration and philosophical introspection. As the landmarks that these writers built, marked, or inscribed became new features in later geographical works, a reciprocal cycle between literature and geography was completed.

Chapter 5 approaches Yuan Zhen and Bai Juyi from a new perspective, building on the previous discussion of the interactions between mid-Tang literature and the contemporaneous advancement of geographic studies. Under the collective denomination "Yuan-Bai," Yuan Zhen and Bai Juyi rapidly ascended to celebrity status in the early ninth century. Their deeply intertwined works attained a vast and penetrating popularity and represented the prevalent literary tastes of their age, especially during the Yuanhe period (806–820). Among other factors, what helped these two poets build up their joint poetic enterprise was a dynamic, interactive geographic dimension in their exchange writing. Over the course of their turbulent migratory careers, they integrated knowledge of routes, landmarks, places, distances, travel speed, and so on into their exchange poems about memories, hopes, dreams, telepathy, and imagined journeys so as to orient each other toward imaginary reunions. In so doing, they tested how geographically unrelated corners of the empire could be connected by the lines of their literary texts and could consequently acquire new meanings in the real world.

I first discuss the respective geographical engagement of Yuan Zhen and Bai Juyi in their individual works. Yuan was an expert in cartography, while Bai had abundant travel experience from an early age. Their different forms of geographic learning resulted in different patterns of interaction between literature and geography in their work. This discussion is meaningful in and of itself, but also serves as a prelude to the poets' much more sophisticated geographical engagement in their long-distance exchange poetry. The chapter then highlights two sets of poems written in two different circumstances: the first pair was written when Yuan was passing Liangzhou (in modern day Shaanxi Province) on his way to Sichuan while Bai remained in the capital. Both poems address the locations of the two friends in a surprisingly coordinated manner, and my analysis indicates that this poetic wonder was made possible only because the two poets were continually tracing each other in their respective mental maps and wrote about this spatial imagination in their poetry. The second two poems are two long-regulated verses (*pailü* 排律) written when Yuan was demoted to Tongzhou in Sichuan and Bai was demoted to Jiangzhou in Jiangxi. Bai's work contains a lengthy section on the local geography of Jiangzhou, while Yuan corresponds with couplets that address one place in the first line and another in the second. This geographic juxtaposition allows Yuan to join and integrate the two otherwise unrelated places in his textual space, enabling him to form connections with his faraway friend. Finally, I study a series of Yuan and Bai's individual and exchanged works that traced the routes of the dissemination of their literary output. As a whole, these works conjured up a textual map for traveling texts and a literary empire dominated by their creations.

This final chapter culminates in a grand finale that weaves the key issues discussed in the previous chapters into the spectacular case of the poetic exchange between the two most popular mid-Tang poets. Their vision of their sphere of influence was based on an understanding of the empire, and their articulation of that vision took the form of a poetic grand map. Their corresponding poems illustrating the rich details of the land and the cultures of Jiangzhou and Tongzhou exemplify the affinity between local geography and local writing, as well as the unique aesthetic texture that such affinity creates. Finally, deep inside the empire's less developed southern frontiers, the two poets did not poeticize the space as merely a metaphorical expression of their banishment, but rather enlivened their respective locations and their friendship through parallel spatial imaginaries boosted by bold formal experiments. Bai and Yuan's lifelong friendship and poetic correspondence amid their constant travel across the Tang territory offer potent examples of the interaction and mutual transformation of geographic advancements and poetic creativity.

Notes

1. See Mitchell, "What Is an Image," in *Iconology: Images, Text, Ideology* (Chicago: The University of Chicago Press, 1986), 7–46.
2. Ibid., 14.
3. Yee, "Cartography in China," *The History of Cartography* (Chicago: The University of Chicago Press, 1994), vol. 2, book 2, 128.
4. Ibid., 127.
5. Ibid., 96–127, and particular 109–110: "...Traditional Chinese cartography is marked by the coexistence of two major tendencies. These two tendencies can be termed measurational (or more broadly, observational) and textualist. The latter strand can be seen as having two aspects: first, a reliance on texts as sources of information in the compiling of maps and, second, a reliance on text to complement the presentation of information in maps."
6. Tian, *Visionary Journeys: Travel Writings from Early Medieval and Nineteenth-Century China* (Cambridge: Harvard University Asia Center, 2012), 22.
7. Ibid., 22.
8. Ibid., 22.
9. Ibid., 4–5.
10. See Yi-fu Tuan, *Space and Place: The Perspective of Experience* (Minneapolis: University of Minnesota Press, 2001; first print 1977), 3–5. Tuan's differentiation of space and place in this book is developed from a similar argument in his earlier article, "Space and Place: Humanistic Perspective," in C. Board, R.J. Chorley, P. Haggett, and D.R. Stoddart, eds., *Progress in Geography*, 1974, vol. 6, 211–252.
11. Tuan, *Space and Place: The Perspective of Experience*, 54.
12. Tim Cresswell, *Place A Short Introduction*, 10. For example, Tim Unwin notices that in Henri Lefebvre's *The Production of Space*, the canonical work on the theory of social space, there is no clear, coherent and consistent differentiation between the notion of space and place. For Lefebvre, space refers to social space, but as Unwin points out, on one level, "The production of space can be seen as espousing the radical thesis that *all* space is produced." Tim Unwin, "A Waste of Space? Towards a Critique of the Social Production of Space," in *Transaction of the Institute of British Geographers*, vol. 25, no. 1 (2000), 11–29, and 26 in particular.

13. Tuan, *Space and Place: The Perspective of Experience*, 6.
14. Cresswell, *Place: A Short Introduction*, 10.
15. For a useful study of this topic with a synthesis of existent scholarship, see Paula Varsano, "Do You See What I See?: Visuality and the Formation of the Chinese Landscape," *Chinese Literature: Essays, Articles, Reviews*, vol. 35, 2013: 31–57.
16. "Site." Def.1 and 2a. http://www2.merriam-webster.com. 2003. Last accessed on May 14, 2017.
17. Brosseau, "Geography's Literature," *Progress in Human Geography*, 18.3 (1994), 334.
18. Ibid., 347.
19. Some of the articles and books addressing this point include: Douglas C.D. Pocock, ed., *Humanistic Geography and Literature (RLE Social & Cultural Geography): Essays on the Experience of Place* (London: Croom Helm, 1981); Brosseau, "Geography's Literature"; Joanne P. Sharp, "Towards a Critical Analysis of Fictive Geographies," *Area* vol. 32, issue 3 (September 2000), 327–34, and others.
20. Bunkse, *Geography and the Art of Life* (Johns Hopkins University Press, 2004).
21. For a review of some early scholarship along this line, see Fabio Lando, "Fact and Fiction: Geography and Literature: A Bibliographic Survey," *GeoJournal* vol. 38, no. 1 (January 1996), 3–18. For recent examples, one can name Robert Tally Jr.'s exploration of the idea of a "literary cartography" in his book *Melville, Mapping, and Globalization: Literary Cartography in the American Baroque Writer* (London: Continuum, 2009).
22. See Hones, *Literary Geographies: Narrative Space in* Let the Great World Spin (New York: Palgrave Macmillan, 2014), 7–8.
23. Liang, "Zhongguo dili dashi lun," *Yinbingshi wenji* 飲冰室文集 (Shanghai: Guangzhi shuju, 1902), vol.14, 19–38.
24. In his article "Main Fields in the Research of Literary Geography," Zou Jianjun lists the following aspects that he believes future studies of literary geography should cover: 1) The influence that writers have received from natural environments; 2) The construction of a geographic space in literary works; 3) The depiction of landscape in literature and its meaning; 4) The emergence of literary schools and their relationship with natural environments; 5) The relationship between literary developments and the change of geographic environment; 6) The influence of geographic discovery on literary texts; 7) Human observation of the cosmic space and its impact on the mindset of literary writers; 8) Differences in

the expressions of geographic spaces by Chinese and Western writers. See Zou, "Wenxue dilixue yanjiu de zhuyao lingyu" 文學地理學研究的主要領域, *Shijie wenxue pinglun* 世界文學評論, 2009.1, 42–45.

25. Zeng, *Wenxue dilixue yanjiu* (Beijing: Shangwu yinshuguan, 2012), the whole book and pp. 12–13 in particular.
26. Yang, *Wenxue dilixue huitong* (Beijing: Zhongguo shehui kexue chubanshe, 2013), the whole book and pp. 32–38 in particular.
27. Liao, *Tangdai de lishi jiyi* 唐代的歷史記憶 (Taipei: Taida chuban zhongxin, 2010).
28. Wang, *Zhongguo gudai shanshui youji yanjiu* 中國古代山水遊記研究 (Beijing: Zhongguo shehui kexue chubanshe, 2008).
29. Hones, *Literary Geographies: Narrative Space in* Let the Great World Spin, 8.

Chapter 1

Geographical Advancements in the Mid-Tang

In Chinese intellectual history, the close affinity between geography and literature did not occur suddenly in the mid-Tang period. The two fields were not clearly demarcated, and therefore texts from both fields had shared thematic and stylistic features for a long time. There were also certain earlier periods in which a rapid advancement in geographical awareness and achievement similar to the mid-Tang can be observed. The interplay between these geographical advancements and contemporary literary developments have been noted by other scholars, and an overview of this historical cross-field affinity will help us better understand its manifestation in the mid-Tang. Moreover, the major geographical works discussed in this overview will reoccur as points of reference in the following chapters.

Next, I will first delineate the heightened geographical awareness shared by a wide range of mid-Tang literati within the contemporary social and historical context. I will also discuss the involvement of major mid-Tang literary masters in geographical study or exploration. I will then introduce Jia Dan's "Map of Chinese and Foreign Lands," situating it in the mid-Tang map culture, along with the map-guides, and Li Jifu's *Maps and Treaties of the Provinces and Counties of the Yuanhe Reign*.

GEOGRAPHY AND LITERATURE BEFORE THE MID-TANG

In Chinese history, the field of geography is most commonly referred to as *dili* 地理, which literally means the patterns/principles of the earth. By the Tang, the notion of *dili* had evolved to an inclusive intellectual rubric through which to observe spatial patterns of nature, customs, politics, and so on.[1] In its inception in archaic times, geographic work, although appearing to delineate patterns in the terrain of the physical world, is in fact best described as a mixture of observation, fantasy, and philosophical conceptualization of inhabited and uninhabited spaces. A fictional journey, a mixture of real and imaginary geographic features, and the imposition of a conceptual world model onto the known world —these narrative and fictional elements are commonly seen in early Chinese geographic works. If we consider narrative and fictionality as key elements of literature, then early Chinese geographical work also exhibits literary quality. With their authors and dates of production untraceable or in debate, it is hard to identify these works' immediate influence on contemporary literary production. Nevertheless, these geographic texts and visual representations became long-lasting sources of reference and inspiration for later intellectual production, especially in terms of a sense of space and an understanding of the world.

"Tribute of Yu" (*Yugong* 禹貢), a chapter in the ancient classic *Book of Documents* (*Shangshu* 尚書), is widely acknowledged as the classical text that marked the inceptive moment of Chinese geography. Likely written in the Warring States period, the work divides the heartland of China into "Nine Provinces" (*jiuzhou* 九州) and organizes the whole world into "Five Zones of Submission" (*wufu* 五服). The Nine Provinces are demarcated along natural components like mountains and rivers and are defined by distinctive geographic features and special local products, such as agricultural products, leather, feathers, and so on. The system of the "Five Zones of Submission," on the other hand, designates the royal or imperial capital as the center of the world and divides the rest of the world into successive concentric zones in which the level of civilization declines as one moves further away from the capital.[2] The nine-province structure establishes the fundamental paradigm for later geographic work. As Joseph Needham observes, "all Chinese geographers worked under its [the "Tribute of Yu"] aegis, drew the titles of their books from it, and tried unceasingly to reconstruct the topography which it contained."[3]

"Tribute of Yu" is an emblematic example of the literary nature of classic Chinese geographic works. It unfolds in a narrative structure, following the movements of the ancient hero Yu along the rivers of China. The entire piece can be seen as a fictional record of how Yu travels across the realm to restore different rivers to their channels and subsequently to recreate China after the flooding.[4] The opening line of the chapter, "Yu divided the land" ("Yu futu" 禹敷土), takes on the typical storytelling style that introduces the protagonist and his action at the beginning of a story.[5] Moreover, although the descriptions of the mountains and rivers seemingly correspond to natural landmarks, in the actual topography of China many of them prove unidentifiable, or are not located where the text shows them. Some river channels are legendary places; others are purely imagined. In other words, part of what "Tribute of Yu" delineates is a fictional geography.[6] Finally, the grid structure of the Nine Provinces is itself an abstract conceptualization of space that was later developed into a complicated philosophical world model composed of a series of grids by the philosopher Zou Yan (c. 250 B.C.).[7]

"Classic of Mountains" (*Shanjing* 山經), another iconic early geographic work that maps out the world along mountain ranges (rather than rivers), is structured as a journey from one mountain to the next. Moreover, the world depicted in "Classic of Mountains" is not human-centered, but rather is inhabited by all manner of monstrous creatures, strange plants, and divine beings. Because of these features, Mark Edward Lewis concludes that the world of the "Classic of Mountains" "is that of the interfaces between the animal, the human, and the divine."[8] According to Lewis, the book is a thorough spatial allocation of myths and legends: "It ["Classics of Mountains"] forged diverse myths and local lore into a systematic account by assigning every creature, divine beings, or ancient story to a place."[9]

Later, additional sections titled "Classics of the Seas" ("Haijing" 海經) and "Classics of the Great Wasteland" ("Dahuangjing" 大荒經) that cover even more remote corners of the land were added to "Classics of Mountains," making an expanded work that is conventionally referred to as *Classics of Mountains and Seas* (*Shanhaijing* 山海經). Throughout Chinese intellectual history, the *Classics of Mountains and Seas* has inspired many artists and literary writers to imagine and represent a mythical and spiritual sphere of the world.

Early maps, although not directly involving literary elements, also contain aspects that would fascinate any literary writer. Early Chinese maps often present real and fictional geography in forms that mix cartographic elements with text, painting, and religious/philosophical charts. In ancient records, maps are referred to as *tu* 圖, which can mean map as well as picture or chart.[10] On the one hand, empirical precision was a key objective of the early cartographers. According to Cordell Yee, in the Qin and Han dynasties, maps were used for military and ritual purposes. They reflected actual locations of geographical features in a mensural manner and were crucial guides for making military plans. They were also presented to the emperor as "part of the ceremonies associated with enfeoffment."[11] On the other hand, it is also possible that ancient maps were used in divination and hieratic activities as symbolic cultural objects.[12] In addition, Yee takes note that some early maps could be approached as works of visual art, as the patterns used on maps as geographic symbols were found to have been engraved on other objects as decorations as well.[13] The tradition of the map as painting and religious chart and with textual annotations continued throughout the premodern period and made maps an important aid for literary and artistic creation.[14]

The four centuries of unity under the Han saw significant developments in geography. The field became better defined, its works attributable to individual authors. Several chapters of Sima Qian's 司馬遷 (145 or 135–90 B.C.) *Records of the Grand Historian* (*Shiji* 史記) offered rich, verifiable information on specific regions and the water system of the empire.[15] Ban Gu 班固 (32–92), the Eastern Han historian, included a "Treatise on Geography" (*dili zhi* 地理志) in his *Book of Han* (*Hanshu* 漢書), thus initiating the tradition of including a dedicated geographic section in official dynastic histories. Han geography, in general, took a big step towards empiricism from the earlier geographic texts. Local customs, products, and the lineage and activities of aristocratic families were common subjects for Han geographers. In addition, regional geography during the Han Dynasty involved a strong historical component. In Ban Gu's "Treatise on Geography," for example, he traces the changes of names, administrative reach, and customs of many cities and regions back into archaic times. The way Han geographers composed regional geography is considered to have established the patterns for the more standardized forms of local geography that came later. These include the map-guides and later the gazetteers (*difangzhi* 地方志).[16]

The abundance of regional geographic data bolstered the development of the rhapsody (*fu* 賦), a major literary genre of the Han and ensuing Wei-Jin periods. Dominating the literary scene for centuries, the rhapsody is usually written in rhymed prose in which an object, feeling, or place is depicted and celebrated in exhaustive detail.[17] Ban Gu himself was a renowned rhapsody writer. His "Rhapsody on the Two Capitals" ("Liangdu fu" 兩都賦) captures the splendor and richness of the two capitals of the Western and the Eastern Han, Chang'an and Luoyang. Scholars notice that Ban Gu's rhapsodies tend toward realistic descriptions of places, and thus are distanced from the excessively lavish style of earlier rhapsody writers.[18] This was only possible because of Ban's impressive command of geographic knowledge. The realistic style of rhapsodies on place persisted into the works of the Eastern Han writer Zhang Heng 張衡 (78–139) and the Western Jin writer Zuo Si 左思 (c. 250 –c. 305). Cordell Yee points out that their rhapsody writing benefited greatly from the geographic training they had received. Zhang Heng was himself an accomplished cartographer and astronomer. His most celebrated rhapsody, "Rhapsody of Two Metropolises" ("Erjing fu" 二京賦), contained rich details about geographic mensuration and surveying. In Zuo Si's "Rhapsody of Three Capitals" ("Sandu fu" 三都賦), the author explicitly acknowledged that he had utilized maps and gazetteers in preparing for the composition of the rhapsody.[19] In short, the rhapsody's detailed depictions of geographic objects were sustained by the contemporary development of regional geography.

The fall of the Eastern Han in 220 and China's subsequent political fragmentation gave rise to a new geographic genre known as "records of the earth" (*diji* 地記).[20] The earliest independent geographic works in Chinese history, they also exerted considerable influence over contemporary and later literature. This geographic genre flourished between the Eastern Han (25–220) and the Southern Song (420–479). Geographic writings prior to this period, such as those by Ban Gu and Sima Qian, were usually part of a larger encyclopedic collection. In the early seventh century, when Wei Zheng 魏徵 (580–643) compiled the *Book of Sui* (*Suishu* 隋書), a bibliographic category of "geography" (*dili* 地理) emerged, and dominating the long bibliographic list were numerous titles of records of the earth. According to David Jonathan Felt, the flourishing of geographical writing in the early medieval period is closely related to contemporary changes in the imperial geography, such as the fragmentation of formally unified territories, the demographic shift southward into the Yangtze River Basin, and so on.[21] Unfortunately, only a small number of records of the earth survived, so later scholars have to rely on the Northern Wei (386–534) geographer Li Daoyuan's 酈道元 (466 or 472–527) *Commentary on the Classic of Waters* (*Shuijing zhu* 水經注) to glean information about what this genre was like. The book contains encyclopedic descriptions of the landscape, history, and culture of regions across China, organized along rivers and streams. In writing the book, Li consulted maps and other geographic works, and incorporated numerous earlier and contemporary records of the earth, both with and without explicit reference to the original texts.[22]

Records of the earth were highly literary in nature. They dealt with natural geographic objects such as mountains and rivers as well as regional products and customs. Felt observes that records of the earth and other similar geographic writing of the time typically move the narrative "through space the way a human might physically travel between places, making observations, and then continuing on to the next location."[23] This is reminiscent of the tradition of very early geographic work, except that in records of the earth the implied traveler is human rather than a legendary figure and the geographic details are much richer and more realistic. For literary scholars, records of the earth are noteworthy for their stylistic descriptions of natural landscape in well-crafted lyrical prose. Wang Liqun argues that records of the earth in early medieval China paved the way for the emergence of the landscape essay in the Tang Dynasty in two important ways: first, it marked a leap in the intellectual awareness of natural landscape in Chinese history; and second, these well-written geographic records were frequently paraphrased in later landscape essays.[24] In fact, Chinese and Western scholars have long found that some of the most celebrated passages in Liu Zongyuan's landscape essays were taken from Li Daoyuan's *Commentary on the Classic of Waters*, which in turn may have been taken from other records of the earth by Li's contemporaries.[25]

This present book is set in the period of grand unification that followed in Chinese history. While my primary focus is the interplay between the new geographic developments of the mid-Tang and contemporaneous literature, it should be noted that writers of the period were by no means oblivious to earlier geographic achievements. Acutely aware of the unprecedented flourishing of geography in their own time, they also understood that this was a culmination of a long process of evolution. The fact of their appreciation of ancient geographic knowledge and works will help us better understand the general significance of geography to mid-Tang intellectuals. Their broad familiarity with geography of the past also afforded them wide latitude to appropriate and deploy previously established terms and ideas in their own very contemporary literary creations.

The mid-Tang writers were exposed to a variety of early geographic advancements. They were familiar with the ideas, images, and narratives in classical geographic texts such as "Tribute of Yu" and *Classics of Mountains and Seas*, or in later geographic monographs by authors such as Ban Gu and Li Daoyuan. In addition, fragments of a wide range of geographic studies by literati geographers from ancient times to the sixth century, including many records of the earth, were transmitted through the numerous encyclopedias, or *leishu* 類書, in the Sui and the Tang. Literally "classified books," *leishu* consists of extracts taken from a variety of earlier writings and classified under different categories. Scholars in general trace the origin of *leishu* back to the early third century, and by the Tang Dynasty, *leishu* had developed into a crucial intellectual resource and literary reference that was widely consulted by the literati when composing literary works.[26]

Understanding the accumulative impact of geographical works on mid-Tang writers is useful for our later discussions, because we sometimes need to tease out different layers of geographical influences in literary texts so as to find out which layer can be attributed to the contemporary geographical development of the mid-Tang period. For example, knowing that the scenic descriptions in Liu Zongyuan's landscape essays draw heavily upon records of the earth from the Six Dynasties, we can then focus on elements in his essays that are not associated with records of the earth and are instead directly related to the geographical changes of his time. Knowing what conventional associations the "Maps of Mountains and Seas" would evoke for Tang poets, we can then look into how the poet Zhang Hu imposed a grand map-view that only became available in his time onto the old map-pictures. Likewise, when we discuss the possible geographical references behind literary writers' engagement with various local spaces of the empire, we need to discern how historical texts from local geography filtered by *leishu* and contemporary local geographic work functioned differently for literary authors. Accordingly, the mid-Tang writers' exposure to earlier achievements in geographical studies such as those introduced previously will be referenced in later chapters as it becomes relevant.

Heightened Geographic Awareness in the Mid-Tang

The mid-Tang period, spanning from the 790s to the 820s, is best described as a time of order in chaos. It was a period when the empire was struggling to recover and reestablish itself in the long aftermath of the An Lushan Rebellion, which ended in 763.[27] Regional wars continued to break out, foreign invasions led to constantly shifting frontiers, and the empire's heartlands were significantly destabilized, becoming sometimes even more insecure than the periphery. On the western border of the empire were the Tibetans.[28] In the northern heartland of the empire, the strong military autonomies in Hebei enjoyed de facto independence and drained the power of the empire from within. The loss of security undermined cultural confidence as well as the deeply rooted imperial rhetoric of superiority. In terms of regaining central control over those problematic regions both militarily and administratively, as well as restoring imperial integrity on a symbolic level, geography emerged as a field of knowledge that was of crucial importance to the empire. As Li Jifu 李吉甫 (758–814), the chief minister and leading cartographer of the time, proclaimed, "To accomplish the present missions and ensure a favorable position in the future, nothing is more pivotal than maps and geography" 成當今之務，樹將來之勢，則莫若版圖地理之為切也.[29]

As a result, a group of leading cultural and political figures of the time, including Jia Dan, Li Jifu, Han Yu, Yuan Zhen, Liu Zongyuan, Liu Yuxi, and Li Deyu 李德裕 (787–850), was seriously involved in producing and studying "maps and geography." Maps and geography constituted a significant part of the mid-Tang intellectual culture, and it was this group of literati that materialized the inner connections between geography and literature and pushed the development of both to an unprecedented level in Chinese history.

The two people who set the advancement of geography in full swing from the 790s to the early 810s were the leading geographers Jia Dan and Li Jifu. The fact of their long-term central positions in the court as chief ministers and their outstanding expertise in geography bespeaks the general social significance of geography in mid-Tang, as well as the importance of geographic competence to a literati's professional life. At court, both Jia Dan and Li Jifu acted as chief organizers of new geographic information acquired from across the empire and beyond imperial borders. Though they had once been occupied with work as regional governors or had suffered from political setbacks, once each was promoted to the position of chief minister, their centrality in the notoriously turbulent and even deadly contemporary political scene became relatively stable, making it possible for them to conduct extensive research with full institutional support. While at court, Jia Dan composed the iconic mid-Tang grand map, the "Map of Chinese and Foreign Lands." For his part, Li Jifu compiled *Maps and Treaties of the Provinces and Counties of the Yuanhe Reign*, the earliest extant collection of local geography that covers the entire Chinese empire. Jia's and Li's emphasis on geography became a shared basic tenet among cultural elites and would be passed down to younger generations.

A decade later, Yuan Zhen's emergence as an important mapmaker further demonstrates the vital role the literati played in the dynamic between political power and advanced geographic knowledge. In 821 Yuan, then a Hanlin scholar responsible for drafting documents for the emperor, submitted to Emperor Muzong several maps that he crafted with his distinguished cartographic and geographic knowledge. The maps tackled several difficult issues that the emperor and the court were facing. For instance, he presented two copies of his four-volume reader-friendly *Map-Guide for Regions to the West and the North of the Capital*

(*Jingxi jingbei tujing* 京西京北圖經). One was for the emperor's own perusal, and the other served as a guide for Princess Taihe's 太和公主 (fl. 821–843) impending journey to marry a Uighur khan. Where the princess should stay at night and where the accompanying troops could find water were—to the embarrassment of the authorities—not easy questions to answer. The route that the princess should take had long been unclear to Tang officials due to the government's diminished territory in recent decades. Yuan Zhen came to the emperor's aid just in time.[30] According to what Yuan told the emperor, he had also finished a "Map of the Western Edge of the Great Tang" ("Sheng Tang xiji tu" 聖唐西極圖), a map of considerable scope that included many administrative regions and neighboring countries in the west.[31] Given the importance of this map, he suggested to the emperor that they have a face-to-face discussion. Yuan thus advanced into a dramatic moment in his political career. The emperor arranged three private meetings with him, which showed an unusual degree of trust and appreciation. In 822, Yuan was appointed a chief minister by the emperor. [32]

Han Yu, too, was an expert in maps and geography, and this expertise also proved to be crucial for his professional achievements. In 811, Han Yu was appointed to the position of the Vice Director of the Bureau of Operations in the Military Department.[33] Han's experience in the bureau gave him a working expertise in maps and military geography. In 817, he was further invited to serve as the military advisor in the Huaixi War that was aimed at a military recovery of the Huaixi area, which had long been out of the control of the central government. During this time, he was known to make accurate suggestions for marching routes.[34] Moreover, a celebratory essay he wrote after the victory in the Huaixi War demonstrates that he was fully aware of the symbolic value of maps in a time of crisis. The prose piece, titled "Stele of the Pacification of Huaixi" ("Ping Huaixi bei" 平淮西碑), highlights a symbolic scene of map reading conducted by Emperor Xianzong 憲宗 (r. 805–820) right after he took the throne:

> After Emperor Xianzong had received his ministers' respects, he studied maps and enumerated the tributary states, saying, 'Alas! Heaven endowed my family with everything and passed it onto me. How can I go to the ancestral shrine without any accomplishments?' All his ministers held these words in awe and busied themselves fulfilling their duties.
>
> 睿聖文武皇帝既受羣臣朝，乃考圖數貢。曰："嗚呼！天既全付予有家，今傳次在予，予不能事事，其何以見于郊廟?" 羣臣震慴，奔走率職。[35]

This scene of the emperor studying a map in front of his subjects marks the crucial moment in which he became determined to recover all the lost territory of the empire. Here, Han evokes the map as a source of political incentive for imperial revival and a symbol of the integrity of the empire.

While Yuan Zhen and Han Yu stood out as well-trained cartographers, Liu Yuxi was known for his contribution to texts of regional geography. In 815, Liu was demoted to Lianzhou, a prefecture in the remote imperial south. A local friend described to him a local natural phenomenon, namely the impressive sea tides, and asked him to write about it. Liu wrote a poem titled "Song for the Pounding Tide with Introduction" ("Tachao ge bing yin" 沓潮歌並引), in whose introduction he made it clear that the poem would be included in the "Record of the Southern Yue" 南越志, a record of local history and geography.[36] In 839, he wrote an essay titled "Record for the New Post Road in the West of the Shannan Circuit" ("Shannan xidao xinxiu yilu ji" 山南西道新修驛路記). At the end of the essay he again revealed that the piece was written at the request of local people to be included in the "Treatise on Geography" ("Dili zhi" 地理志) in the official record of history.[37] At the time, it was quite common for compilers of local geographic works to include relevant literary texts in geography, but Liu's explicit mentioning of such inclusion demonstrates that as he was writing, he was highly aware that he was composing literature for geographical works. Later chapters will show that Liu's literary writing on places in general resembles geographic writing in many ways.

In addition to making maps and writing geographic texts, literati writing itself established new landmarks in the newly developed imperial frontiers, which in turn became new favorites of future geographic works. Accompanying the mass immigration to the south during and following the Rebellion, leading cultural figures including Han Yu, Liu Zongyuan, Bai Juyi, Liu Yuxi, and Yuan Zhen were all sent to the south for official posts, mostly in the form of political demotions. While in the south, many of them actively discovered new landscapes, built new buildings, and designed new cities and dwelling places. They wrote to record their explorations or to mark them in the physical terrain by inscribing the literary texts into the landscape. In this sense, their relocation was part of the broader geographic changes of the empire, and they actively participated in these changes as both political officials and literary writers. The landmarks they created can be found in geographic works from as early as the Late Tang, and some still persist today. This is yet another way that the mid-Tang literary writers contributed to contemporary and future geographic advancement, and it required a high level of both geographic and literary sensitivity.

As a result, the mid-Tang witnessed a number of achievements in geographic exploration that were among the most important during the dynasty and even in the medieval period as a whole. Next, I would like to introduce two major geographic achievements in the mid-Tang, both of which are closely related to our discussion of mid-Tang literature. They are the grand maps and the map-guides.

Maps and Grand Maps in the Mid-Tang

Jia Dan's "Map of Chinese and Foreign Lands" provided mid-Tang writers with fresh perspectives with which to imagine and navigate the world through literature. To prepare for our investigation into the intellectual conversation between maps and mid-Tang literature, in this section I will introduce how Tang maps represented space in general, and more specifically, how Jia's grand map reconfigured the vast space of the world and the empire in unprecedented ways.

Tang maps carried forward many characteristics of traditional Chinese cartography discussed earlier. Presenting geographic features of the earth, they were used as navigational and organizational tools for military and political purposes.[38] Evidence shows that precision was crucial to a map's value. For instance, in Yuan Zhen's 821 "Memorial Presenting the 'Map of the Western and the Northern Frontiers'" ("Jin xibei biantu zhuang," 進西北邊圖狀), he informed the emperor that his map would surpass all others of the same kind in its geographic precision. After decrying previous maps submitted to the Emperor for not being able to survive the test of field inspection, Yuan boasts with full confidence that his map "records every tiny detail and is precise down to feet and inches" (*Xianhao bizai, chicun wuyi* 纖毫必載, 尺寸無遺). He even suggests that the emperor summon veteran military officers to prove his work's precision.[39] Elsewhere, the testimony was provided by third-party commentators, as can be seen in the following contemporary evaluation of the "Map of Chinese and Foreign Lands" in *Tang huiyao* 唐會要 (*Institutional History of the Tang*): "Someone consulted the map and asked others in his prefecture about its accuracy. They all said it was accurate and without error" (*You pitu yi wenqi junren zhe, jiede qishi, wu xuci yan* 有披圖以問其郡人者，皆得其實，無虛詞焉).[40] Unfortunately, because of the limited records on Tang cartography, we know very little about exactly what technologies of surveying and measurement these Tang mapmakers developed to achieve such precision.[41]

At the same time, Tang maps were integral to a larger enterprise of political, artistic, and religious expressions of space, and entailed active interactions with other representational modes such as landscape painting, calligraphy, and poetry. As we have discussed, the Chinese word *tu* had long meant picture, map, or religious chart. It also carries the meaning of "design," "plan," "plot," or "enterprise," and can be used as a verb to refer to planning, anticipating, giving thought to, or dealing with something.[42] In the following Tang example quoted from Li Deyu's memorial to Emperor Wuzhong (r. 840–846) accompanying his geographic work, we see how maps and governance are correlated through their shared character "*tu*":

> I, your subject, have been in western Sichuan for a short period. Here I have considered the local advantages and disadvantages, and have taken control of some strategic places, as part of the *enterprise* of governing the empire's distant lands. For this purpose, I wrote the thirteen-volume *Xi'nan beibian lu* ("Record of the Southwestern Frontier Defenses"). I also drew small *maps* of the towns I established, where rice, salt, and armor are all well in supply. When Xiao He took over the *maps* and documents of the Qin Dynasty, he knew all the strategic points. Concerning the affairs of the state and the army, nothing is more crucial than this. [italics added.]
>
> 臣頃在西川，講求利病，頗收要害之地，實盡經遠之圖*。因著《西南備邊錄》十三卷。臣所創立城鎮，兼畫小圖*，米鹽器甲，無不該備。昔蕭何收秦圖*書，具知阨塞，軍國之政，莫切於斯。[43]

In this memorial, the first *tu* 圖 means "(the imperial) enterprise," and the phrase *jingyuan zhi tu* 經遠之圖 means "the enterprise of governing distant lands." The next two *tu* refer to "map." Bringing in the maps of the Qin Dynasty, the author emphasizes their crucial function in conveying key information about the political and military affairs of the state. This *tu* therefore corresponds well with the first *tu* meaning the imperial enterprise. The juxtaposition of the word *tu* with its different connotations enhances the shared epistemological underpinning and points to the significance of maps to imperial governance.

Tang maps continued to share an affinity with paintings as well. Given that the Late Tang art historian Zhang Yanyuan 張彥遠 (c. 815–877) included maps in the catalogue of paintings in his *Famous Paintings through History* (*Lidai minghua ji*, 歷代名畫記), Yee claims that maps must have constituted a genre of painting during the Tang.[44] In fact, according to art historians such as Pan Gongkai and Zhang Hongxing, from the Six Dynasties to the Tang, landscape painting as a genre developed out of the tradition of cartography.[45] This fluid relationship between map and painting allowed a map reader to appreciate a map within the traditions of landscape painting, or vice versa.

Moreover, just like ancient maps, Tang maps used numerous textual explanations either on the map or accompanying the map in the form of an appendix, connecting maps closely to intellectual writing. The most notable case is Jia Dan's "Map of Chinese and Foreign Lands," which was accompanied by an encyclopedia of forty volumes titled "Notes on Administrative Jurisdictions and Foreigners in Four Directions" ("Gujin junguo xiandao siyi shu" 古今郡國縣道四夷述). This prose work was intended to explain in great detail what was represented on the map, and it constituted a collection of Jia's work over several decades. Since both the map and the encyclopedia were highly inclusive, they functioned together as a visual and textual interface designed to accumulate and integrate a wide range of knowledge.

The category of grand map, or *datu* 大圖, refers to maps of the empire or world maps in premodern China. It is also sometimes called imperial maps, or *guotu* 國圖.[46] Today we know of two grand maps of the mid-Tang. One is Jia Dan's "Map of Chinese and Foreign Lands" and the other is "Map of the Territorial Record" ("Dizhi tu" 地志圖) by Li Gai (dates unknown). These maps joined landscape with seascape, situated the empire in the cosmos, and established order between the Chinese and the non-Chinese. Of the two, only Jia's map was an official project sponsored by the court; it marked the highest cartographic achievement in the Tang Dynasty and was highly informative and inspirational for the production of an image of the world in mid-Tang and Late-Tang literature. Li's map, on the other hand, was independently made and was much less known than Jia's map. Today, we can only read about his map through Lü Wen's 呂溫 (772–811) "Preface to the Map of the Territorial Record" ("Dizhi tu xu" 地志圖序). Our focus here will therefore be on Jia's map.

Jia Dan was commissioned by Emperor Dezong's court in 784 to draw an imperial map, after a rebellion that had threatened to bring down the empire.[47] In his early career Jia had worked in the Foreign Affairs Department (*honglu si* 鴻臚寺) and used the opportunity to interview diplomats from neighboring countries and tributary states. He subsequently developed an extensive body of knowledge of foreign cultures and societies as well as geographic connections between the Tang Empire and foreign countries.[48] As a result, he was said to be able to "distinguish, illustrate, and point out the origins and developments of all the plains and treacherous regions both within the country and without, and all the barbaric minorities' aboriginal customs" (*Jiuzhou zhi yixian, baiman zhi tusu, qufen zhihua, beijiu yuanliu* 九州之夷險，百蠻之土俗，區分指畫，備究源流).[49] Still, it took him more than a decade to actually draft the map and the accompanying encyclopedia-like map explanations. The map took its final form in 801, and Jia died four years later.

Although Jia's map, like all other Tang maps, is no longer available today, a twelfth-century map carved on a stele, titled the "Map of Chinese and Foreign Lands" ("Huayi tu" 華夷圖), is considered by historians to be a later, simplified version of Jia's map. The following is a transferred image of the stele:

Figure 1. The twelfth-century "Map of Chinese and Foreign Lands," stone rubbing, 1903 (now in the Library of Congress, USA).

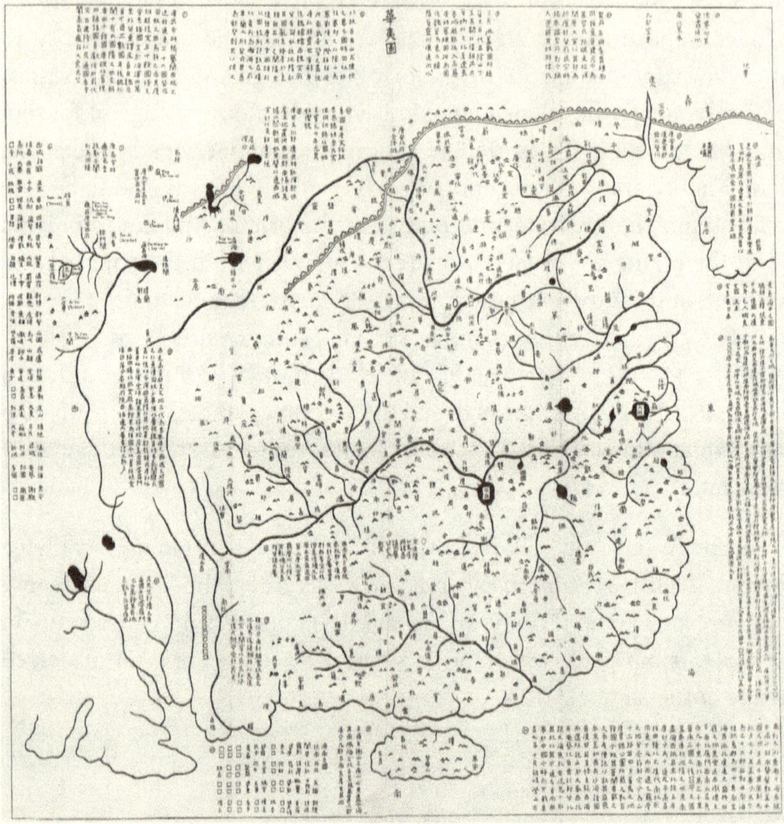

What first impressed viewers about Jia's map was its sheer scale. The simplified version is already nearly three square feet and contains about 500 place names. Yet as Richard Smith points out, it "records only the most familiar names from among the 'several hundred countries' listed by Jia Dan on his own world map."[50] The original map is allegedly much larger. Jia claims that his map was 30 feet long and 33 feet high, constructed on a scale of 1 inch to 100 *li.* Joseph Needham estimates that "it must therefore have covered an oikoumene of 30,000 *li* from east to west and 33,000 from north to south. . . and must therefore had been, to all intents and purposes, a map of Asia."[51]

In this map, a fine order is imposed on both the cultural and the natural world. Like most premodern Chinese grand maps, it takes the world model of "Tribute of Yu" as its guiding principle, but buttresses it with considerable newly acquired geographical knowledge. In Jia Dan's memorials to Emperor Dezong that accompany his map, he describes the map as one that "pushes outward to the four seas, and divides the nine provinces within" (*Qi datu waibo sihai, neibie jiuzhou* 其大圖外薄四海，内別九州).[52] "Four seas" here refers to the four territorial directions, and so covered in Jia's map is the broadest imperial territory with a finer demarcation of nine provinces in the heartland. Beyond the nine provinces, this map made visually accessible a transcultural and ethnically varied world by recording and classifying "ten thousand countries laid out like chess pieces" (*Wanguo qibu* 萬國棋布) and "a hundred barbarian regions weaving a colorful embroidery" (*Baiman xiucuo* 百蠻繡錯).[53] As Hilde De Weerdt says, "In place of a map of dynastic territory alone, Jia Dan had drawn a map that showed and listed more states than any other Chinese map would identify for centuries to come."[54] Jia's ultimate aim in drawing the map was "to distinguish the civilized from the barbaric and determine the locations of high mountains and grand rivers; to narrow the world down on silk and delineate hundreds of prefectures in the painting" (*Bie zhangfu zuoren, dian gaoshan dachuan; suo siji yu xiangao, fen baijun yu zuohui* 別章甫左衽，奠高山大川；縮四極於纖縞，分百郡於作繪).[55]

Although the conception of the Chinese and the foreign—or *hua* and *yi*—as separate groups had appeared in China long before Jia's era, Jia was the first in the history to visualize it in the form of an imperial map. There are many ways to translate *hua* and *yi* into English, each with different cultural and political implications. In this book, I translate *huayi* mostly as *the Chinese* and *the foreign*. As I have explained earlier, the category of *yi* as it is represented in Jia's grand map is first and foremost a geographic notion. It includes the vast regions beyond the Chinese heartland, ranging from Chinese frontiers populated by non-Han ethnicities to regions outside the Tang empire that were connected to the Tang through diplomatic relationships. Consequently, *hua* may refer to the Han-oriented heartland within the empire, or more broadly the Chinese land of different ethnicities within the empire as distinct from other geographical regions in the world. While it is hard to find singular English words that correspond to these two nebulous and shifiting concepts, "the Chinese" and "the foreign" best capture the geographical and cultural divergences between *hua* and *yi*, as well as the alienness of *yi*. In the context of specific literary texts, where it is certain that *hua* and *yi* are evoked as ethnic markers, I translate these terms as the Han and the non-Han.[56] The translation that I avoid intentionally is that of "the Chinese" and "the barbarian." There are occasions where the dyad of *hua* and *yi* is employed to highlight the superiority of the Chinese and the inferiority of the foreign in contemporaneous intellectual work, but in the maps and map-related texts that I deal with in this book, this is rarely the case.

In Jia's map, the grand visualization of the *hua* and the *yi* in one picture emphasizes the all-inclusiveness of the empire in the cultural, political, and, most importantly, geographical senses more than the differences between the two categories. In a time of imperial restoration, a grand map showing a clear imperial order naturally became a symbol of imperial vitality. This symbolic value can be seen in many literary works that we will discuss in the following chapters. For now, I want to bring in a record from the Late Tang period. After the rebellion lead by Huang Chao 黄巢 (878–884), when the downfall of the dynasty seemed to be only a matter of time, the poet Sikong Tu 司空圖 (837–908) wrote a "Record of the 'Map of Chinese and Foreign Lands'" ("'Huayi tu' ji" 華夷圖記), in which he considered the map to be the embodiment of the past imperial prosperity: "Among what has been left in the ashes, there is still Chief Administrator Jia's records of the imperial territory. Open his map and examine its details, and we can understand the causes of victories and failures" (*Weijin suocan, shangcun Jia Puye Dan fangyu zhizhi, pitu jiaoyan, chengbai kezhi* 煨燼所殘，尚存賈僕射耽方域之志，披圖校驗，成敗可知).[57]

It is not clear to us today how the map was copied and circulated in the country, but evidence for its dissemination is seen in poems and stories of the Late Tang and the Five Dynasties. The historian Xin Deyong suggests that one way of making the map more accessible to the general public might have been to copy it on the walls of public spaces such as famous Buddhist temples.[58] Representing a milestone in Chinese cartographic history, Jia's grand map had profound influence on later cartographical work in China. As Richard Smith confirms, a close resemblance to Jia's map can be found in most extant "world maps" in the Song Dynasty.[59]

Map-Guides in the Mid-Tang

In addition to the grand map, another genre of mid-Tang geography that had intricate interactions with contemporary literature is the map-guides, or *tujing*. Important mid-Tang writers such as Han Yu, Liu Yuxi, Yuan Zhen, and Bai Juyi all mentioned the map-guide in their writing, a phenomenon unheard of in earlier times. As the most authoritative encyclopedia of local knowledge compiled by local scholars and officials, map-guides provided literary writers with abundant geographical materials to engage with in their writings concerning the local spaces.[60]

A map-guide consists of maps (*tu*) and detailed explanatory texts (*jing*) on a given local region. This kind of work can be dated back to at least as early as the Eastern Han, and it developed as a minor geographic genre throughout early medieval times.[61] It was not until late medieval times that the map-guide was increasingly integrated into the imperial administration of local places, either as an important reference for centralized projects on local geography, such as *Maps and Treaties of the Provinces and Counties of the Yuanhe Reign*, or as a primary site guide for local officials. Unlike ordinary geographic works produced by individual intellectuals, the map-guides were regularly updated to subsume historical and contemporary information from existing sources and surveys, and therefore they became the most comprehensive source of local knowledge. The Sui court required that all administrative regions "categorize their customs, products, and maps, and send them to the State Affairs Department" (*Tiao qi fengsu wuchan ditu, shangyu Shangshu* 條其風俗物產地圖，上于尚書).[62] Following this practice, the Tang government requested updated regional map-guides every three or five years, and map-guides from across the country were collected and organized in the Bureau of Operations of the central government.[63]

Several reasons may have contributed to the growing importance of the map-guides in the Sui-Tang period. Xin Deyong attributes it to institutional change. In 583, the Sui changed the traditional practice. In the past, local assisting officials were appointed by regional administrative heads, but under the new system, the central government appointed all the officials and reappointed them to new posts every four years. In principle, the positions were held by people from other prefectures. This practice continued through the Tang. As a result, the newly appointed officials, who were outsiders to the region, needed help to be able to systematically familiarize themselves with the local geography. Xin believes the best way to do this was to require prefects and counties to compile map-guides that comprehensively recorded the local geography.[64] In addition to the institutional change of appointing outsiders as local officials, as Hua Linfu points out, there might have been many other impetuses behind the increase in the production of the map-guides. Hua draws on a broad range of historical examples of the importance of the map-guides to the social life, governmental management, and literary compositions of the Tang. These examples range from the management of hydraulic works to changes in the names of regions, from official and personal travels to the emperor's consultations, from military campaigns to the writing of biographies.[65]

Since all of the maps and most of the texts of the map-guides are now lost, what we now know about them is largely gleaned from two sources, which represents a small fraction of what the Tang literati could see of the map-guides in their own day. The first resource is fragments of two map-guides from a place called Shazhou, which were compiled at different times. They are the *Map-Guide of Shazhou* (*Shazhou tujing* 沙州圖經), compiled between 670 and 710, and the *Map-Guide of the Shazhou Military Commandery* (*Shazhou dudufu tujing* 沙州都督府圖經), compiled at the turn of the eighth century and amended in 766. They were both discovered in Dunhuang in the twentieth century. These fragments have drawn interest from many important modern historians, and were fully annotated by Li Zhengyu 李正宇 in the 1990s.[66] The second resource consists of allusions to the map-guide found in various historical and literary sources, especially the Song compilations that contained a great number of texts written during the Tang and earlier periods.[67] In 2007, the historian Hua Linfu 華林甫 compiled and organized 548 allusions to various map-guides dating from the Sui and Tang through the Five Dynasties, significantly expanding our knowledge of the map-guides produced across a wide range of areas in the empire.[68] With the Dunhuang fragments and these allusions, we are now in a better position to study the map-guide, its function in the imperial political system, and its interactions with Tang literature.

The map-guides organized a broad range of subject matters that covered a cross-section of local knowledge. In the extant fragment of the *Map-Guide of the Shazhou Military Commandery*, we find twenty-four categories of local features:

> Sources of water, ditches, moats, marshes, weirs, old dikes, grand halls, saline and alkaline lands, salt lakes, lakes, post houses, prefectural schools, county schools, medical schools, altars of gods of the land and agriculture, miscellaneous gods, the strange and the anomalies, temples, tombs, halls, "earth-rivers,"[69] ancient cities, auspicious signs, poems and ballads.

> 水、渠、壕塹水、澤、堰、故堤、殿、鹹鹵、鹽池水、泊、驛、州學、縣學、醫學、社稷壇、雜神、異怪、廟、塚、堂、土河、古城、祥瑞、歌謠。[70]

These categories generally consisted of classified landmarks, except for "the strange and the anomalous" and "auspicious signs," which were records of local events and phenomena, and "poems and ballads," which consisted of collected literary texts. The descriptions of landmarks in the *Map-Guide of the Shazhou Military Commandery* generally used the prefectural city as the geographic center for measurement of distance, and then provided records of each entry's geographic position, characteristic features, and sometimes founding history.[71]

In addition to existing landmarks, the *Map-Guide of the Shaozhou Military Commandery* includes a report of twenty-two categories that the Shazhou area reportedly lacked, such as animal husbandry offices (*jianmu* 監牧), the "loose control" prefectures of ethnic minorities (*jimi zhou* 羈縻州), the Four Rivers (the Yellow River, the Yangzi River, the Huai River, and the Ji River) (*Jiang He Huai Ji* 江河淮濟), chaste women and women martyrs (*jiefu lie'nü* 節婦烈女), and so on.[72] This list of missing features suggests that the Tang government might have standardized a classificatory system to be used by all of the compilers of map-guides. Yet without any further explanation, the list of the missing categories is puzzling. As Li Zhengyu points out, the report of some missing categories contradicts the information the same map-guide provides.[73] We might conjecture that this fragment of the *Map-Guide of the Shazhou Military Commandery* was not derived from a final version to be submitted to the central government, but from a work in progress that contained obvious mistakes. Whether or not this is the case, this list reflects certain practical difficulties in fitting a large amount of geospatial data and historical references into standardized categories, which the surveyors of the local land would have had to tackle.[74]

As well-crafted examples of writing, some texts in the map-guides, especially those associated with geological wonders or local legends, manifest a high level of literary composition. In the *Map-Guide of the Shazhou Military Commandery*, for example, there is a vivid description in parallel prose of a famous sand dune, the Mingsha Mountain (*Mingsha shan* 鳴沙山) in Dunhuang:

> The mountain moves incessantly, and its peaks are not fixed in position. Its deep valleys can burst suddenly into hills, and the high rocky hills turn into valleys. Some of its peaks are so steep they seem chopped by blades, and a lone summit rises as though carved out. At night one suspects that there is no mountain here, but in the morning the mountain reaches the sky. There is a wellspring in the mountain that the flowing sand does not cover. When horses gallop or men walk over the sand, the sound is like thunder.
>
> 其山流動無定，峰岫不恒。俄然深谷為陵，高崖為谷，或峰危似削，孤岫如畫。夕疑無地，朝已幹霄。中有井泉，沙至不掩。馬馳人踐，其聲若雷。[75]

This description of a moving, shape-shifting sand dune incoporates an ancient trope: "Deep valleys can burst suddenly into hills, and the high rocky hills turn into valleys." This is a reworking of two lines in the poem "Alignment in the Tenth Month" ("Shiyue zhijiao" 十月之交) in *The Book of Songs* (*Shijing* 詩經), "High ledges became deep valleys, and valleys rose up into ridges" (*Gaoan weigu, shengu weiling* 高岸為谷，深谷為陵).[76] In the original context of the poem, the dramatic geological changes are caused by a tremendous earthquake in the Zhou dynasty.[77] In this passage, the writer ingenuously brings the old trope of natural disaster into the new geographical scene with a sense of wonder. The ever-changing character of nature, the painterly scene, and certain significant details such as the wellspring and the thunderous noises all give the description a strong literary force comparable to that of the landscape essays composed by known masters of literature.

The *Map-Guide of Jiujiang* (*Jiujiang tujing* 九江圖經) provides an even more intriguing example of the mélange of geographical description and curious local history, indicating that the map-guides were important sources of local legends and anecdotes:

> The Ganquan ["sweet spring"] water source is to the south of the county's Ganquan Post-House. Its water is sweet, and its fragrance lingers after it is drunk, hence its name. The mountain is accordingly named Ganquan Mountain. According to the prefectural *tujing*, "In the past a rudder was carried down by the creek flowing from the summit. The native people therefore named the creek 'Rudder Down Creek.' When Huan Yi was prefect of Jiujiang, he would often give his attendants provisions and send them on trips deep into the mountains, expecting they would find unusual things. Once they reached a place with a large lake and a decaying boat beside it. Huan Yi had been suspicious when he first heard about the rudder being carried down from the summit. He was convinced when he heard that there was a boat."
>
> 甘泉水在縣南甘泉驛之南，其水味甘，飲訖猶有餘香，因以名焉，其山即曰甘泉山。按《州圖經》云："昔山頂有船柂，從頂沿流而下，土人亦名為柂下溪。桓伊為九江刺史，常遣左右齎糧尋山之奧，冀睹非常，乃至一處，見有大湖，湖側有敗船，當時聞有柂流下，甚疑惑，後聞有船，方驗。"[78]

This passage from a map-guide quotes from an even earlier map-guide, telling the story of how a local official finds proof for a local legend.[79] This framing structure of a map-guide within a map-guide and a legend within an anecdote gives the record a flavor of intricate storytelling. And yet all the stories told here are immediately related to a local place name, the "Rudder Down Creek," and are therefore a legitimate part of local geography.

Some texts in the map-guides made their way into compilations that feature works with literary value. For example, the Song Dynasty compilation of *Extensive Records of the Taiping Era* (*Taiping guangji* 太平廣記 completed in 978) is known as the most important source of early fiction available today.[80] In it, twenty stories were taken from twelve different map-guides, with six from the *Map-Guide of Shezhou* (*Shezhou tujing* 歙州圖經, c. 806) from the mid-Tang. These twenty entries are mostly legends and fantastic stories related to local landmarks, and the longer ones run to several hundred characters. In *Extensive Records of the Taiping Era*, these stories were dispersed under categories such as "dragons," "aquatic species," "birds," "plants," "marvelous monks," "mountains," and so on. Literary compilations such as these would make stories from the map-guides part of the literary tradition, to be shared by numerous talented contemporary and future literary authors who would otherwise never have come in contact with the original map-guides.

Li Jifu's Maps and Treaties of the Prefects and Counties of the Yuanhe Reign

The earliest extant geographic study of all the administrative regions of the Chinese empire, Li Jifu's *Maps and Treaties of the Prefects and Counties of the Yuanhe Reign*, is one of the most significant geographical achievements in premodern China. Although the work did not command as much literary interest at the time as Jia Dan's grand map and the map-guides, it stands as a testimony to the heightened geographic awareness that swept across elite communities. In the following, I will discuss the features of this work as they pertain to contemporaneous spatial consciousness and analyze why the magnificent geographic compilation held less sway among literary writers as compared to other contemporary geographical works.

Completed in 813, Li's work aimed to provide a comprehensive and accurate geographic guide for Tang rulers to efficiently govern the empire as they formulated military strategies, managed population and taxation, and attempted to recentralize power. The work was originally comprised of forty volumes, with maps for each military commandery. The original maps have all been lost, but most of the texts are still extant today.[81] In the preface, Li elucidates his view on the desired type of geographical study: as mentioned, he argues that "to accomplish the present missions and ensure a favorable position in the future, nothing is more pivotal than maps and geography." In particular, he stresses the need to acquire accurate and up-to-date geographical information as it related to administrative and military matters. Throughout the work, he takes careful note of the locations of natural barriers and military fortresses, the demographic statistics of each administrative unit, and local products and tributes. He also utilizes the cartographical system of "Four Extremes and Eight Extents" (*sizhi badao* 四至八到) to provide the distance between a given place and its surrounding referential points in up to sixteen directions. This system, which we will revisit in detail in chapter 2, allows geographers and viewers to accurately pinpoint positions in the empire in relation to their neighboring cities/prefects, as well as to the two capitals Chang'an and Luoyang. In addition to stressing utilitarian values, Li's work also had a symbolic dimension. He chose to include West Sichuan in the last volume, although the region had already fallen to the Tufan. This inclusion, while pragmatically useful, also shows how mid-Tang geographers sought to pioneer efforts to restore the Tang order and imperial territories in symbolic ways.

Li's masterpiece relied on local geographical works for its success, including map-guides, as they supplied regularly updated local data. However, he also demonstrated substantial disagreement with authors of these works about what good geography entailed.[82] His criticism cited in the following quote, although not explicitly directed toward the map-guides, points to a crucial difference between state-sponsored imperial geographical projects and local geographical works, including the map-guides:

> Moreover, several thousand ancients and contemporaries discussed geography. Those who admire the ancient times often gather materials from the past but ignore the present. Those who collect folktales and customs usually relate the questionable and lose sight of the factual. They embellish the prefects and states, and chronicle individual lives; they seek out tombs and gravesites

> to verify ghosts and spirits; they drift into heresies and none of this touches upon the fundamental or essential. With regard to the hills, lands, mountains, and rivers, and the advantageous and disadvantageous positions for military offense and defense, this is all rooted in geography, but they do not write about them. How are they going to assist the enlightened ruler in controlling vital positions in the realm, and in commanding the lives of the people? To reap the benefits of preserving territories and favorable positions, and to show ways to master terrains and govern lands, this is what your humble minister has carefully studied, and is offered for the sagacious sovereign's perusal.
>
> 況古今言地理者凡數十家，尚古遠者或搜古而略今，采謠俗者多傳疑而失實。飾州邦而敘人物，因丘墓而徵鬼神，流於異端，莫切根要。至於丘壤山川，攻守利害，本於地理者，皆略而不書，將何以佐明王扼天下之吭，制羣生之命？收地保勢勝之利，示形束壤制之端，此微臣之所以精研，聖后之所宜周覽也。[83]

As is clearly stated in this passage, for Li Jifu, the only meaningful geographical information is contemporary geographical data for administrative and military use. Many categories included in the local map-guides, such as local personalities and ancient legends, are considered unreliable and insignificant. This means, in Li's view, that the map-guides devote too much attention to "heresies." Such heresies, however, were precisely what fed the imagination of the local writers of the time. It is for these reasons that Li's work, emphatically utilitarian and oblivious to the imaginative, proved to be less relevant in the mid-Tang literary world. Literary irrelevance notwithstanding, the gigantic work still demonstrates the eagerness with which the mid-Tang elites pursued new, accurate, and comprehensive geographic information. This not only points to the mindset of the generation of geographers who shouldered the lofty burden of imperial rejuvenation, but also bespeaks the overall prominence of ideas about space in the mid-Tang collective consciousness. This emphasis on space, variously manifested in individual works, lays the ground for further analysis.

CONCLUSION

This chapter provides a general, contextualized picture of modes of interplay between geography and literature during the mid-Tang. At different moments in Chinese history, geographic knowledge filtered into literature in different ways. In the mid-Tang, when concern about wars and the aspiration for territorial restoration dominated the intellectual scene, a group of cultural figures emerged who were highly accomplished in both geography and literature. These dual talents allowed them to integrate different types of contemporary geographical knowledge into their poetic writings, in particular in the "Map of Chinese and Foreign Lands" by Jia Dan, the map-guides, and *Maps and Treaties of the Prefects and Counties of the Yuanhe Reign* by Li Jifu. While the all-inclusive grand map represented a vast, ambitious world image centered around the empire, map-guides expressed the depth of the empire through minute descriptions of numerous local details. Li Jifu's *Maps and Treaties of the Prefects and Counties of the Yuanhe Reign*, although holding less sway among literary writers compared to the other two types of geographical work, nevertheless elucidated some basic knowledge and underlying systems that were commonly employed in contemporaneous geographical studies. These three types of geographical work provided inspiration for literary writers and led them in different directions, thus allowing us to explore various modes of interplay between the two disciplines.

Notes

1. For a detailed review of the conceptual history of *dili*, see David Jonathan Felt, "Patterns of the Earth: Writing Geography in Early Medieval China" (Ph.D. diss, Stanford University, 2014), Chapter One, 49–145.
2. For more about the "Tribute of Yu," see Mark Edward Lewis, *The Flood Myths of Early China* (Albany: State University of New York Press, 2006), 30–35.
3. Needham, *Science and Civilization in China* (Cambridge; New York: Cambridge University Press, 1954–2008), vol. 3, 503.
4. Mark Edward Lewis, *The Construction of Space in Early China* (Albany: State University of New York Press, 2006), 249.
5. For English translation, see James Legge, ed. and trans., *The Chinese Classics* (London, 1865) vol. III, pt. 1, 92.
6. Wang Chengzu 王成組, *Zhongguo dilixue shi: Xianqin zhi Mingdai* 中國地理學史: 先秦至明代 (Beijing: Shangwu yinshuguan, 2005), 18–19.
7. For more discussions on Zou Yan's methods, see Lewis, *The Construction of Space in Early China*, 249–251.
8. Ibid., 285.
9. Ibid. 285.
10. For a detailed discussion of the etymology of *tu* in archaic writing in China, see Nathan Sivin and Gari Ledyard, "Introduction to East Asian Cartography," *The History of Cartography*, vol. 2, book 2, 26.
11. Yee, "Cartography in China," 77.
12. Ibid., 72.
13. Ibid., 147–150.
14. Yee has made it clear that the difficulty of separating cartographic from artistic traditions "is not confined to early artifacts" but is applicable even after the two traditions "matured." Ibid., 151.
15. Chapters rich in geographical information in *Records of the Grand Historian* include: "Treatise on Rivers and Canals" (*Hequ shu* 河渠書); "Biographies of Commodity Producers" (*Huozhi liezhuan* 貨殖列傳); "Biographies of Dayuan" (*Dayuan liezhuan* 大宛列傳), an account of geographies and histories of polities in the empire's western frontier; "Biographies of Southwestern Barbarians (*Xinanyi liezhuan* 西南夷列傳), among others.

16. Hou Renzhi 侯仁之, *Zhongguo dili xue jianshi* 中國古代地理學簡史 (Beijing: Kexue chubanshe, 1962), 14–16. Local map-guides that first appeared in the Eastern Han Dynasty and flourished in the Sui and Tang Dynasty were commonly considered the prototypes of gazetteers. See Hua, "Suitang tujing jikao shang," 142–143.
17. Cao Daoheng and David R. Knechtges, "Han *Fu* 賦 (*Fu* of the Han)." In David R. Knechtges and Taiping Chang eds., *Ancient and Early Medieval Chinese Literature: A Reference Guide, Part One* (Leiden: Brill, 2010), 317–333.
18. Ibid., 323.
19. Yee, "Cartography in China," 132–133.
20. The genre was also called *dizh*i 地志 or *diliji* 地理記, see Felt, "Patterns of the Earth: Writing Geography in Early Medieval China," 118–120.
21. Ibid., 7.
22. Wang Liqun, *Zhongguo gudai shanshui youji yanjiu*, 64–86.
23. Felt, "Patterns of the Earth: Writing Geography in Early Medieval China," 6.
24. Wang, *Zhongguo gudai shanshui youji yanjiu*, 80–88.
25. As William Nienhauser says, "Liu Zongyuan's description of several rivers in his 'Dialogue with the Stupid Creek' ("Yuxi dui") is similar to that of Li's and suggests he may have been familiar with Li's commentary." See "*Landscape Essays*," in *Nienhauser* et al., *Liu Tsung-y ü an*, (New York: Twayne Publishers Inc., 1973), 68. Inspired by the Ming Dynasty scholar Yang Shen's 楊慎 (1488–1559) commentary, Wang Liqun provides an interesting case of some vivid descriptions of geographic features recycled in early medieval records of the earth such as Yuan Shansong's 袁山松 (?–401) "Record of *Yidu's Mountains and Rivers*" 宜都山川記, Li's *Commentary on the Classic of Waters*, and Liu's essay "To the Little Rock Pond West of the Little Hill" 至小丘西小石潭記. See Wang, *Zhongguo gudai shanshui youji yanjiu*, 83–84.
26. For a recent comprehensive study of this topic, see Xiaofei Tian, "Literary Learning: Encyclopedias and Epitomes," in *The Oxford Handbook of Classical Chinese Literature (1000 BCE–900 CE)*, eds. Wiebke Denecke, Wai-Yee Li, Xiaofei Tian (New York: Oxford University Press, 2017), 132–146.
27. As Nicolas Tackett argues in his book *The Destruction of Chinese Medieval Aristocracy*, "Whereas historians for a thousand years have tended to portray the post-An Lushan Tang as a long period of gradual dynastic decline," the dynasty in fact "remained relatively stable until

quite late in the ninth century," 148. For a broader overview of the An Lushan rebellion and its impact on the empire, see *The Cambridge History of China, vol.3: Sui and T'ang China, Part 1*, Denis Twitchett ed. (New York: Cambridge University Press, 1979), 464–560.

28. In the winter of 763, the Tufan [or Tubo] 吐蕃 army (Tufan being a medieval Tibetan kingdom) conquered and sacked Chang'an, occupying it for several weeks. According to the Tibetologist Sam van Schaik, this event "set the tone for a new phase of Sino-Tibetan relations. While the Chinese annalists continued to write as if their emperor was lord over all neighbouring 'barbarians', the Chinese now had to treat the Tibetans as equals, and reluctantly agree to treaties placing the Sino-Tibetan border only a few hundred miles from their capital. Though Chang'an remained in Chinese hands, the Tibetans camped frighteningly close by and attacked almost every autumn, the traditional nomadic campaigning season. [...] One after another, the western cities, China's gateway to the Silk Route, fell to the Tibetan army." Schaik, *Tibet: A History* (New Haven: Yale University Press, 2013), 27–28.
29. Li, *Yuanhe junxian tuzhi* 元和郡縣圖志 (Beijing: Zhonghua shuju, 1983), 2.
30. *Yuan Zhen ji bianjian jianzhu: sanwen juan* 元稹集編年箋注: 散文卷, Yang Jun 楊軍 ed. (Xi'an: Sanqin chubanshe, 2008), 710–711.
31. Ibid., 707.
32. Zhou Xianglu 周相錄, *Yuan Zhen nianpu xinbian* 元稹年譜新編 (Shanghai: Shanghai guji chubanshe, 2004), 202. Yuan's service as chief minister was disastrously short. His geographic acumen might well have helped him get there, but it was not sufficient to keep him in place.
33. Lü Dafang 呂大防, et. al., *Han Yu nianpu* 韓愈年譜 (Beijing: Zhonghua shuju, 1991), 11.
34. Ibid.,12.
35. See *Han Yu wenji huijiao jianzhu* 韓愈文集匯校箋注, Liu Zhenlun 劉真倫and Yue Zhen 岳珍 eds., (Beijing: Zhonghua shuju, 2010), 10.2195.
36. Liu, *Liu Yuxi quanji biannian jiaozhu* 劉禹錫全集編年校注, eds. Tao Min 陶敏 and Tao Hongyu 陶紅雨 (Changsha: Yuelu shushe, 2003), 4.222.
37. Ibid., 19.1265.
38. Many Tang maps specified their represented geographic scope in the title, as we can see in examples such as the *Shidao tu* 十道圖 ("Map of the Ten Administrative Regions") by Li Jifu, the *Xiyu tu* 西域圖 ("Map of Western Regions") by Wang Zhongsi 王忠嗣 (706–750), *Sheng Tang xiji*

tu 聖唐西極圖 ("Map of the Western Edge of the Great Tang") by Yuan Zhen, and *Zhudao xingcheng xuemai tu* 諸道行程血脈圖 ("Map of the Routes of All the Administrative Regions Formed Like Blood Vessels") by Ma Jingshi 馬敬寔 (dates unknown). Most of the titles of Tang maps were recorded in "Treatise on Bibliography" ("Jingji Zhi" 經籍志) of the *Jiu Tang shu* 舊唐書 (*Old History of the Tang Dynasty*) and "Treatise on Arts and Letters" ("Yiwen Zhi" 藝文志) of the *Xin Tangshu* 新唐書 (*New History of the Tang Dynasty*). They were not accorded a category but were mixed in with various other geographic writings. For an overview of Tang geographic writings and maps, see Shi Nianhai 史念海, Tangdai lishi dili yanjiu 唐代歷史地理研究 (Beijing: Zhongguo shehui kexue chubanshe, 1998), 2–9. Other titles of Tang maps are found in individual mapmakers' and commentators' writings, in two of the most notable cases, by Yuan Zhen and Lü Wen.

39. Yuan, *Yuan Zhen ji bianjian jianjiao: sanwen juan*, 707–708.
40. Wang Pu 王溥, *Tang hui yao* (Beijing: Zhonghua shuju, 1955), 36.659–660.
41. There is too little information for scholars to trace the scientific progress Tang mapmakers made. The Tang maps fell into a wide gap of over almost a millennium between two monumental works on mapmaking. The first was the "six principles of mapmaking," or *zhitu liufa* 制圖六法, developed by Pei Xiu 裴秀 (224–271) of the Western Jin Dynasty (265–316), which was regarded as the beginning of Chinese scientific or mathematic mapmaking. The six principles were used to measure different topographical features, such as mountains, rivers, and roads, and to define their relations. The six principles were recorded in Pei Xiu's biography in the *Jinshu*. Here I use Smith's translation and brief explanation of them: "Proportional measure or graduated divisions (*fenlü* 分率), the determination of map scale; 2) regulated view (*zhunwang* 準望), the depiction of the correct relationship between the different parts of a given map; 3) road measurement (*daoli* 道里), a way of fixing distances based on triangulation; 4) levelling (or lowering) of heights (*gaoxia* 高下), presumably a system of ground measurement; 5) determination of diagonal distance (*fangxie* 方邪), to be used in depicting mountains and rivers; and 6) straight-ending of curves (*yuzhi* 迂直), apparently a measuring technique for undulating surfaces." Smith, Chinese Maps, 25–26. The second work that demonstrates cartographical technology after Pei's treatise was the "Map of the Tracks of Yu" (*Yuji tu* 禹跡圖), carved on a stele in 1136, which shows the first known appearance of

the Chinese cartographic grid. For more discussion of this map, see Yee, "Cartography in China," 47. Admittedly the gap between the two cartographic works is too wide to allow a linear history of the scientific progress of Tang mapmakers. Given this gap in the evidence, Yee says, "We should remember that, in an important sense, the history of Chinese cartography has yet to be written." Yee, "Cartography in China," 35.

42. For a discussion of *tu*'s application as both a verb and a noun, see Yee, "Cartography in China," 72.
43. Li Deyu, "Memorial to Present the *Xi'nan beibian lu*" 進《西南備邊錄》狀, in *Li Deyu wenji jianjiao* 李德裕文集校箋, eds. Fu Xuancong 傅璇琮 and Zhou Jianguo 周建國 (Shijiazhuang: Hebei jiaoyu chubanshe, 2000), 18.351.
44. Yee, "Cartography in China," 137. Along the lines of Yee's argument, Craig Clunas further discusses the historical process from the late Ming to the early eighteenth century, in which a "dissociation of sensibility" developed between maps and paintings. See Clunas, *Pictures and Visuality in Early Modern China* (London: Reaktion Books Ltd., 1997), 81.
45. See Pan Sheng, *Ditu de zuozhe jiqi yuedu: yi Song-Ming wei hexin de zhishishi kaocha*, 11–12.
46. According to the *Beitang shuchao* 北堂書鈔 (*Excerpts from Books in the Northern Hall*), the third century *Fangzhang tu* 方丈圖 ("One Zhang-Square Map") by Pei Xiu was a scaled-down version of an "Old great map of all-under-heaven" 舊天下大圖. See Yu Shinan 虞世南, comp., *Beitang shuchao* (Tianjin: Tianjin guji chubanshe, 1988), 96.404. Jia Dan used both "grand map" and "imperial map" to refer to his maps in his memorials to Emperor Dezong: "The grand map pushes outward to the four seas, and divides the nine provinces within [the empire]" 其大圖外薄四海，內別九州; "In the first year of the Xingyuan Reign (784), I humbly served you and was commissioned to make the imperial map" 去興元元年，伏奉進止，令臣修撰國圖. See Liu Xu 劉昫 et al., *Jiu Tang shu*, in *Ershisi shi suoyinben* (Beijing: Zhonghua shuju, 1975), 138.3786.
47. This rebellion is commonly referred to by historians as *Jingyuan zhiluan* 涇原之亂 ("The Rebellion of Jingyuan") or *Fengtian zhinian* 奉天之難 ("The Catastrophe of Fengtian"). After Emperor Dezong took the throne in 779, he pushed with increasing aggression to reclaim the court's power over Hebei. In the autumn of 783, amidst a series of hard-fought and interlocking campaigns against the military commissioners of Hebei, a mutinous army attacked Chang'an, forcing Emperor Dezong to flee to Fengtian. What followed was a series of pitched battles between

the government army, rebels, and some precarious pro-government military forces. In the summer of 784, he returned to the capital and resumed his reign, but the regions of Hebei continued to enjoy full autonomy. For a contemporary account of the rebellion, see Zhao Yuanyi 趙元一, *Fengtian lu* 奉天錄 (Beijing: Zhonghua shuju, 1985). See also *Jiu Tang shu*, 12.337–345.

48. For studies of Jia's work on the routes that connected the Tang with the Islamic world, see Hyunhee Park, *Mapping the Chinese and Islamic Worlds: Cross-cultural Exchange in Pre-modern Asia* (Cambridge: Cambridge University Press, 2012), 29–34.
49. *Jiu Tang shu*, 138.3784.
50. See Smith, *Chinese Maps*, 27–28. From the Qing Dynasty up to the present time, there has been extensive scholarship in various languages on the stele map and its relationship with Jia Dan. Xin Deyong's article "Shuo fuchang shike Yuji tu yu Huayi tu" 說阜昌石刻《禹跡圖》與《華夷圖》 arguably represents the most advanced study on this topic, not only drawing on important previous scholarship but also providing many refreshing new interpretations and hypotheses. See *Yanjing xuebao* 燕京學報 (Beijing: Beijing University Press), vol. 28, 2010: 1–72.
51. As Joseph Needham says, "It must therefore have covered an oikoumene of 30,000 *li* from east to west and 33,000 from north to south. . . and must therefore had been, to all intents and purposes, a map of Asia." Needham, *Science and Civilization in China*, vol. 3, 543–544.
52. See *Jiu Tang shu*, 138.3784.
53. Ibid., 138.3785.
54. De Weerdt, "Maps and Memory: Readings of Cartography in Twelfth- and Thirteenth-Century Song China," in *Imago Mundi*, vol. 61, no. 2, 2009, 154.
55. *Jiu Tang shu*, 138.3786.
56. In his groundbreaking work *Ethnic Identity in Tang China*, Marc Abramson finds that Han and non-Han ethnic identities were explicitly expressed and differentiated during the Tang period. With regard to *hua* and *yi*, while many scholars consider these to be notions representing cultural rather than ethnic differentiations, Abramson believes they had real ethnic connotations and insists they be translated as the Han and the non-Han. Nevertheless, he admits that cultural and political themes continuously found their ways into Tang ethnic discourses, especially during the mid-Tang period, which is the temporal framework of this study. In other words, while the *hua* people essentially refered to the Han

people, and *hua* culture to the Han-oriented culture, non-Han ethnicities occupied disparate positions in the Tang ethnic discourse depending on whether they were assimilated into the *hua* culture and were loyal to the *hua* empire or not. In this sense, in different circumstances, the term *hua* may as well include peoples and regions that were not necessarily Han but have been well assimilated to it. See Mark Abramson, *Ethnic Identity in Tang China* (Philadelphia: University of Pennsylvania Press, 2008), the whole book and xxi in particular.

57. *Quan Tang wen* 全唐文 (Zhonghua shuju, 1983), 808.8493.
58. Xin, "Shuo fuchang shike Yuji tu yu Huayi tu," 34.
59. Smith, *Chinese Maps: Images of "All Under Heaven"* (New York: Oxford University Press, 1996), 28.
60. Although we cannot identify who the surveyors were in most cases, there is evidence that literati appointed as local officials participated in the compilation of the map-guides. As Liu Yuxi recounts in his "Record of the Official Hall of the Prefect of Kuizhou" 夔州刺史廳壁記, the high-Tang politician Yuan Qianyao 源乾耀 (?–731) was noted for contributing substantially to the local map-guides while serving as the prefectural adjutant. Given that Yuan lived roughly a century before Liu's time, Liu's acknowledgment suggests that the government possibly retained records of how the local map-guides were compiled and amended through the ages. See *Liu Yuxi quanji biannian jiaozhu*, 16.1064–1065.
61. Hua Linfu 華林甫, "Sui-Tang tujing jikao shang" 隋唐圖經輯考上, *Guoli zhengzhi daxue lishi xuebao* 國立政治大學歷史學報, vol. 27, 2007.5, 143.
62. Wei Zheng 魏征 et.al., *Suishu* 隋書 (Beijing: Zhonghua shuju, 1973), 33.988.
63. Xin Deyong 辛德勇, "Tangdai de dilixue" 唐代的地理學, in *Tangdai diyu jiegou yu yunzuo kongjian* 唐代地域結構與運作空間, ed. Li Xiaocong 李孝聰 (Shanghai: Shanghai cishu chubanshe, 2003), 442–443.
64. Xin Deyong 辛德勇, "Tangdai de dilixue" 唐代的地理學, 442–443.
65. Hua, "Sui-Tang tujing jikao shang," 144–145.
66. Li, *Guben Dunhuang xiangtuzhi bazhong jianzheng* (Taibei: Xinwenfeng chubanshe, 1998) and *Guben Dunhuang xiangtuzhi bazhong jianzheng* (Lanzhou: Gansu renmin chubanshe, 2008). In this chapter I cite the edition published by Gansu renmin chubanshe. Earlier scholars on the materials include Luo Zhenyu 羅振玉 (1866–1940), Paul Pelliot (1878–1945), Wang Chongmin 王重民 (1903–1975), Zhou Shaoliang 周紹良

(1917–2005), and Ikeda On 池田温 (1931). For their studies, see Luo, *Mingsha shishi yishu zhengxubian* 鳴沙石室佚書正續編 (Beijing: Beijing tushu chubanshe, 2004), 34–36. Pelliot, "Le 'Cha-tcheou-tou-fou-t'ou-king' et la colonie sogdienne de la region du Lob Nor," *Journal Asiatique*, 1916: 111–23. Wang, *Dunhuang guji xulu* 敦煌古籍敘錄 (Beijing: Zhonghua shuju, 1979), 113–117. Zhou, "Du Shazhou tujing juan zi" 讀《沙州圖經》卷子, in *Dunhuang yanjiu* 敦煌研究, vol. 2, 1987: 27–33.

67. Such records can be found in, for example, *Finest Flowers from the World of Letters* (*Wenyuan yinghua* 文苑英華), *Materials Read by the Emperor in the Taiping Xingguo Era* (*Taiping yulan* 太平御覽), *Imperial Geography of the Taiping Xingguo Era* (*Taiping huanyu ji* 太平寰宇記), and *Extensive Records Assembled in the Taiping Xingguo Era* (*Taiping guangji* 太平廣記).
68. Hua, "Sui-Tang tujing jikao shang" *Guoli zhengzhi daxue lishi xuebao* vol. 27, 2007.5: 141–213, and "Sui-Tang tujing jikao xia" 隋唐圖經輯考下, *Guoli zhengzhi daxue lishi xuebao*, vol. 28, 2007.11: 1–92.
69. "Earth-rivers" refers to an ancient construction used for military defense and reconnaissance. It was dug out of the road, two *zhang* in width and two *chi* in depth, and covered by fine sand and dust. Inspectors swept and checked it every day to record the footprints of men and horses that entered the region. See Li, *Guben Dunhuang xiangtuzhi bazhong jianzheng*, 112.
70. Li, *Guben Dunhuang xiangtuzhi bazhong jianzheng*, 45–58.
71. To provide references to the development of some historical landmarks, the *Shaozhou duduful tujing* alludes to historical accounts such as the *Hanshu* 漢書, the *Xiliang lu* 西涼錄, the *Houwei shu* 後魏書 and the *Shiliu guo chunqiu* 十六國春秋.
72. Here is the full list of the missing categories: Animal husbandry offices, the "loose control" prefectures of ethnic minorities, the Four Rivers (the Yellow River, the Yangzi River, the Huai River, and the Ji River), seas, irrigation canals, prefectural and county towns, checkpoints on rivers, major mountains and rivers, iron production, steles, famous people, loyal subjects and filial sons, chaste women and women martyrs, fortresses, mausoleums and tombs, platforms and pavilions, post pavilions, mine caves, places graced by emperors, places visited by famous ministers and generals, military agro-colonies. 監牧、羈縻州、江河淮濟、海溝、陂、宮、郡縣城、關鑵津濟、嶽瀆、鐵、碑碣、名人、忠臣孝子、節婦烈女、營壘、陵

墓、臺榭、郵亭、礦窟、帝王遊幸、名臣將所至、屯田. See Li, *Guben Dunhuang xiangtuzhi bazhong jianzheng*, 55.

73. Ibid., 126.
74. Based on all the extant texts of the *tujing*, Hua Linfu has reconstructed a list of categories, mostly in more general terms, in an attempt to reflect the broad subject range of Sui and Tang *tujing*: "The development of administrative regions, locations of governmental offices, borders, distances (including those by river), household registration, ancient sites, ancient kingdoms, ruins, palaces, cities, mountains, mountain ranges, hills, mountain ridges, valleys, stone chambers, fortresses, sources of water, creeks, ponds, river fords, deep pools, ditches, sandbars, marshland, swamps, lakes, seas, rapids, shoals, bays, dikes, wells, springs, temples, villas, parks, pavilions, platforms, gates, products, transportation, historical figures, women martyrs, lands conferred for nobility, legends, ethnic peoples, mausoleums, tombs, sepulchers, shrines, ancestral temples, associated counties." 政區沿革、治所、境界、里距(含水路)、戶口、古跡、古國、廢墟、宮、城、山、嶺、丘、岡、谷、石室、關、水、溪、池、津、潭、溝、洲、澤、淀、湖、海、瀨、灘、灣、陂、井、泉、寺、園、苑、亭、臺、門、物產、交通、人物、烈女、封爵、傳說、民族、陵、墓、塚、祠、廟、屬縣。Hua, "Sui Tang tujing jikao shang," 148.
75. Li, *Guben Dunhuang xiangtuzhi bazhong jianzheng*, 45.
76. See *The Book of Songs: The Ancient Chinese Classic of Poetry*, trans. Arthur Waley, edited with additional translations by Joseph R. Allen (New York: Grove Press, 1996), 172. *Shijing zhuxi* 詩經注析, Jiang Jianyuan 蔣見元and Cheng Junying 程俊英 eds. (Beijing: Zhonghua shuju, 2005), 574.
77. See *Shijing zhuxi*, 575.
78. Hua, "Sui Tang tujing jikao shang," 176–177.
79. According to Hua, the map-guide being quoted in the record is probably compiled before the Sui. The record itself is likely from the map-guide compiled during the Sui.
80. William Nienhauser, edit., *The Indiana Companion to Tradtional Chinese Literature* (Bloomington: Indiana University Press, 1986), 745.
81. For a comprehensive study of Li and his work, see Qu Xinfu 屈新福, *Li Jifu jiqi Yuanhe junxian tuzhi* 李吉甫及其《元和郡縣志》 (M.A. thesis, Anhui University, 2009).
82. According to Hua Linfu, in compiling *Maps and Treaties of the Provinces and Counties of the Yuanhe Reign*, Li Jifu referred to the map-guides as

a reference for regional geography. Hua, "Sui Tang tujing jikao shang," 149.

83. Li, *Yuanhe junxian tuzhi*, 2.

Chapter 2

The Big Picture

Poetic Visions and the Cartographic Eye

This chapter explores the creation of some spectacular images of the world in mid-Tang literature through a cross-media investigation that crystalizes the crucial link between literature and maps. In the mid-Tang, these world images were summoned across media and circulated in both public and private venues, offering audiences a new set of geographical knowledge and a revitalized view of the cosmos. While literary representations of the world are a well-trodden subject in the field of literary history, this chapter introduces a cartographic dimension, which has been less studied in its own right and which is rarely considered in its relation to literature. Reading some selected examples of mid-Tang literary works about the world alongside contemporaneous developments in mapmaking, this chapter will offer a new interpretation of some mid-Tang literary authors' poetic reconfigurations of the world.

On the literary side, the early ninth century saw some intriguing new ways of representing the world. Li He's "Dreaming Heaven" imagines how the Nine Provinces 九州 of China emerge as misty, watery, and small when gazed upon from the moon. Liu Zongyuan's poems and essays evoke geographic details typically seen in maps to negotiate with imperial power from the peripheries. In "Two Poems on Contemplating a Painting of Mountains and Seas" ("Guan shanghai tu ershou" 觀山海圖二首), the poet Zhang Hu looks at a well-known ancient landscape painting as though reading an imperial map. All of these examples imply a powerful, clairvoyant observer of the world, and signify the magnitude of the contemporary literati's large-scale spatial imagination.

These poetic wonders corresponded to a concurrent development in the making of maps, and grand maps in particular. In the mid-Tang, makers of grand maps claimed imperial authority over the world by inscribing boundaries, both political and cultural, onto the map. In particular, Jia Dan's "Map of Chinese and Foreign Lands" represented the highest achievement of mapmaking among the literati of the time. It combined new cartographic methods with artistic expressions and rendered painterly images of terrestrial features in both cosmic and historical order. The conversation between grand maps, exemplified by Jia's map, and the representative world images in contemporary literary works forms the basis of this chapter.

The chapter approaches the relationship between mid-Tang maps and poetic imagination as a "conversation." On the one hand, maps served as visual references to aid the cosmic visions of literary authors and enabled new ways of representing landscapes and travel in poetic writing. On the other hand, poetic imagination animated cartographic representations and afforded them fresh expressive power. Maps and literature coactivated each other and coproduced a powerful image of the world. Like all conversations, the one between maps and literature evinced both ease and tension. Poetic thinking could endorse the image of the world represented in a map, but could also subtly manipulate or challenge the image, as well as the imperial ideology behind it. Such interplay gave rise to a multilayered spatial aesthetics and offered an extended interpretative space.

Methodologically, this chapter will not be a simple juxtaposition of maps alongside poems, insofar as no Tang maps survive in their original form today. Rather, my primary reading strategy is to capture traces of the conversation between maps and literary expression through carefully chosen examples of literary works. What does literature tell us about the grand maps? How do literary works represent or mirror a map reading experience? How are map-informed inspirations transformed into poetic aesthetics in literature? Finally, how does literature stretch or twist the world views represented in maps for its own expressive purposes? In short, my goal is not to make a simple list of literary mentions of maps, but rather to recognize the map-viewing eye and to understand its remarkable effects in the intriguing world images produced in literature.

In a broader sense, the cross-media conversation that resulted in a renewed representation of the world did not only take place between literature and maps. Other representational modes, for instance landscape painting, also participated in the intermedia world configuration in medieval times. Indeed, the production and reception of landscape paintings shared similar patterns and dynamics with the ways maps and literature were produced and read. My discussion will therefore involve spatial imaginaries in media beyond literature and maps as they become relevant. In this sense, using cartography as the primary reference point not only deepens our understanding of the imagination and representation of the world in mid-Tang literature, but also provides insights into the making of world images on a multimedia horizon.

The rest of the chapter will be devoted to close readings of four literary authors' works in the light of their exposure to maps. I will first introduce a poem by Du Fu on his map viewing experience. Although the poem about a regional painting-map was written during the transitional period between the High Tang and the Mid Tang, it anticipates the map-informed poetic imaginary in mid-Tang literature in many ways. I will then focus on Li He's poem "Dreaming Heaven," a masterpiece of Tang poetry. The poem's imaginative view of the earth and its sense of grand historical change have impressed many critical readers. I will reexamine the world image Li created in relation to the advancement of cartographic skills as shown in Jia Dan's grand map. Following Li He's poetic visualization of the cosmos is a discussion of selected poems and essays by Liu Zongyuan. In these works, Liu twisted the ways the relative locations of frontier posts are represented on the map, thus transforming mere place names and locations into expressive icons with rich political meaning. In so doing, he developed a renewed sense of empowerment in his conversation with the imperial center from a marginalized position on the frontier. Finally, we will examine Zhang Hu's two poems about a "Painting of the Mountains and Seas." In the poems, Zhang imposed a grand map view of the world onto his viewing experience of a landscape painting. While earlier poets tended to read the painting as a tool for religious meditation, Zhang's cartographically inspired reading of the painting pushed him into an artistic realm that embraced an imperial vision. My discussions of these poets in the following sections will not only trace the rich conversation between maps and literature, but also open up important topics that the following chapters will more fully explore, such as representations of localities, migration and travel, and long-distance communication.

DU FU AND THE POETICIZATION OF A MAP-READING EXPERIENCE

Before focusing on the mid-Tang, I would like to introduce a poem about reading a painting-map, written by Du Fu. Titled "At a Banquet Lord Yan Held in His Hall, We All Composed Poems on a Painting of the Cliff-Paths of Sichuan, and I Was Assigned the Rhyme Word 'Kong,'" the poem manifests the versatility of Tang maps in general, and at the same time exemplifies how multiple aspects of a map can produce a strong visual impact on its reader and generate a poetic aesthetics in the setting of a public gathering:

> The sun overlooked the quiet hall.
> Densely painted, the picture of the land was sublime.
> Sword Pavilion to the north of Starry Bridge,

Pine Prefecture to the east of Snowy Peak.
The mountains were not divided into Chinese and foreign,
Rivers connected Wu and Shu.
My excitement joined the glowing mists and rose-pink clouds,
How fortunate that my pure cup was never empty.

日臨公館靜
畫滿地圖雄
劍閣星橋北
松州雪嶺東
華夷山不斷
吳蜀水相通
興與烟霞會
清樽幸不空[1]

This poem was composed near the end of the An Lushan Rebellion in 762, when Du Fu was in Chengdu, Sichuan. As the title shows, a painting-map of the famous Sichuan cliff-paths was shown to a group of literati in a banquet held in the official hall of Lord Yan (Yan Wu 嚴武 726–765), then the Military Commissioner of the Western Sichuan Circuit (*jiannan jiedushi* 劍南節度使) and Du Fu's patron. Sichuan is a basin region shielded from the tumults of the outside world by forbidding mountain ranges. To reach the affluent plains of Chengdu, travelers would have to endure the perils of the cliff-paths, that is, the narrow plank roads wedged into mountain ridges as pathways through the insurmountable natural barriers. The challenges and austere beauty of the path had long been a source of literary inspiration.[2] Yet it is only on the painting-map that a comprehensive view of many routes and places is vividly visualized in both words and graphic displays.[3] Thus, at Yan's banquet, the guests were invited to play a literary game around the table by using the same rhyme word to poeticize their viewing experiences.

As the first line suggests, the painting evokes wonder and leads to a moment of quiet viewing. There is a nod to the formulaic celebrations of the surroundings that generally begin a banquet poem, but attention is immediately redirected toward the painting in question, saving considerable space for the poet to transform his sight into insight. The word *ditu* 地圖 (lit. picture of the land) in the second line captures the close affinity between a map and a landscape painting in medieval times. At work in Du Fu's poem, however, is not a two-way interaction between map and painting, but a triangulated relationship of co-inspiration among geography, visual art, and poetry.

The second couplet organizes geographic spaces and navigates among four well-chosen Sichuan landmarks: Sword Pavilion, Starry Bridge, Pine Prefecture and Snowy Peak. These place names are arranged into counterpointed parallelisms, creating an added layer of sensual associations with the viewing experience: the radiance shared by *jian* 劍 (sword) and *xing* 星 (star) and the seasonal chill created by *song* 松 (pine trees) and *xue* 雪 (snow). These visual and aural effects and their associations both capture an atmospheric sharpness of the natural beauty of these places and impart an evocative wintriness to the painting.

The third couplet broadens the geographic scope. The poet's vision is clearly not limited by the painting, but rather extends in spatial continuity beyond its regional confines. "Chinese and foreign" (*huayi* 華夷), the imperial conception of difference, is lost in the ever-extending mountain ranges. Wu and Shu, two historical regions hundreds of miles apart, are linked by the surging Yangtze River. Du Fu often invokes images of Wu and Shu, two of the three kingdoms locked in feuding after the fall of the Han Empire. The river does more here than connect separate macroregions. It also reconciles a historical antagonism, just as the mountains bridge ethnic difference. Curiously, Sichuan's cliff-paths, prominently featured in the painting's title, are conspicuously absent in the poem. This is perhaps because the metaphorical effect of the cliff-paths, which denote both geographical demarcation and spatial connection, is already richly present in the visual images the poem constructs, rendering a specific invocation of the icon unnecessary.

In the concluding couplet, the first line elevates the viewing experience to the level of glowing mists and clouds, adding a hint of transcendence.

Then, in the second line, the poet's concentration on his map reading/poetry writing is interrupted by the filling of the wine cup. The reader's attention is thus shifted toward the poet paying his respects to the host. The wine cup brings the poet back to the social order of the banquet while reinforcing the enjoyment of reading the painting-map in a communal setting. We should remember this triangular correspondence among the map, the poem, and the wine cup as exhibited in Du Fu's work, for we will encounter it again next in Li He's "Dreaming Heaven."

The poem, merely an occasional composition about a regional painting-map, nevertheless demonstrates the poet's greater vision of the imperial realm and his sharp sense of the geospatial organization of the wartime empire. In fact, this poem is not the only example of Du Fu recounting his encounters with maps during the Rebellion. Wartime perils forced officials and travelers to resort to maps for the geographical knowledge that had become crucial for the embattled empire and its roving subjects alike. When Du Fu took a minor military position in Huazhou 華州 (present-day Hua County in Shaanxi Province) in 758, he drafted a memorial for the prefectural intendant about the presentation of a military operations briefing map to the court.[4] In 759, he gave up the post and moved with his family to Qinzhou 秦州 (present-day Tianshui County in Gansu Province), a relatively peaceful area where he could seek solace among his relatives and friends. While there, he composed twenty poems, later grouped together as "Unclassified Poems from Qinzhou" ("Qinzhou zashi" 秦州雜詩).[5] In the third poem of that collection, he related his utilization of a map of Qinzhou.[6]

In this sense, the map that Du Fu encountered at Yan's banquet was but one of several maps that he engaged with during the Rebellion. This context allows us to further read Du's poem as a reflection on his own wartime experience as mediated by the map. Du fled from the empire's heartland to his refuge in Sichuan along the same rivers and mountain routes as are shown on the map. Now, in 762, as the war was drawing to an end, he may well have been envisioning a return to his northern hometown through the reading of the map as well. While the Rebellion had thrown the empire into chaos and disintegration, the poet could nevertheless conjure up imagery of reunification in his poem, in the line "The mountains were not divided into Chinese and foreign, / Rivers connected Wu and Shu." His physical traversing of the different realms of the empire on his way into the basin inspired his poetic mind to perceive beyond the scope of the map and the antagonism of the war, and to envision a more connected world.

Composed and performed in a gathering both official and communal, the poem also addresses a collective experience shared by its war-torn audience. The Rebellion rattled the whole empire, sending its ruling classes on a massive exodus into the hinterland. This had the consequence of connecting the Sichuan basin ever more intimately with the outside world.[7] Following the emperor who temporarily took refuge in Sichuan when the Rebellion first broke out, many court officials and literati sought shelter in Sichuan. When some of them were appointed to new posts in Sichuan, they began enthusiastically holding poetry events with relocated and local literati, especially after the political situation was more or less settled. Yan Wu, who accompanied Emperor Xuanzong 玄宗 (r. 712–756) on his relocation to Sichuan in 756, was one of these officials. Many of the guests at Yan's banquet may have also shared similar experiences of wartime relocation. Therefore, when Du Fu composed and presented his poem at the banquet, it in effect mediated the map into a register of the pain, loss, survival, and hope of the audience.

To conclude, the painting-map of Sichuan's cliff-paths becomes a medium that evokes Du Fu's poetic contemplation. The encounter between poetry and map starts at the moment when the excitement of reading a map / enjoying a painting is combined with that of composing a poem. The cartographic eye moves on as the poem progresses; the painterly representation of geographic features with annotation of place names is transformed into a poetic arrangement of images, associations, and creative dynamism. This poetic contemplation does not belong to the poet alone, but is rather part of a social event and resonates with a collective wartime experience. Such is the complete process of the poeticization of map viewing as represented in Du Fu's poem.

The poem predated Jia Dan's grand map "Map of Chinese and Foreign Lands" in 801, but it already showed the poet's ambition to move beyond the representational limits that a map as a cartographic device prescribed and the geographic boundaries that the regional map demarcated. Entering the ninth century, with the new cartographical innovations of Jia's grand map, the scope and depth of the view available in maps were greatly expanded. Consequently, the ongoing conversation between maps and literature was also reinforced and refreshed.

Li He and a Cosmic View of the Grand Map

In Chinese literary history, the mid-Tang poet Li He is known for his wild imagination across a broad temporal and spatial span.[8] One of his most acclaimed poems was "Dreaming Heaven," which depicts an image of the world from the viewpoint of the moon. The impulse to represent or to "see" the human world from without may be eternal and universal. Seldom, however, do we encounter a medieval poem like "Dreaming Heaven" that so vividly imagines the human realm as a radically contracted, colorful micro-world.

The first half of the poem delineates the celestial world on the moon in detail, while the second half presents images of the earth as if they were captured from an extraterritorial perspective. The fantastic imagination displayed in the second half of the poem has long impressed its readers.[9] The poet and critic Li Jian 黎簡 (1747–1799) believed that in the otherworldly quality of poem's conclusion, Li He even surpasses the great High Tang poet Li Bai 李白 (701–762).[10] Despite the high acclaim the poem has garnered, we do not know in which year during the poet's short life the poem was composed, nor do we know the context of the composition. The lack of biographical context may reinforce many critics' belief that the poem is a result of an unparalleled personal genius. Yet some of the visual patterns embodied in the poem are not wholly a personal invention, but they can be interpreted fruitfully in the context of mid-Tang map culture. First, here is the poem:

> The old hare and the shivering toad wept sky-blue tears,
> The cloud tower half opened with slant and white walls.
> The jade wheels creaked upon the dew, dampening their auras,
> One meets with Sumburgh-bells and girdle-gems on the
>   cassia-scented path.
> Yellow dust and clear water below the three mountains,
> A thousand years changed like galloping horses.
> Gazing afar toward the middle land that appears
>   like nine dots of mist,
> All the seawater filling a cup.[11]

> 老兔寒蟾泣天色
> 雲樓半開壁斜白
> 玉輪軋露濕團光
> 鸞佩相逢桂香陌
> 黃塵清水三山下
> 更變千年如走馬
> 遙望齊州九點煙
> 一泓海水杯中瀉[12]

Within the Chinese poetic tradition, the first half of the poem can be linked to *youxian shi* 遊仙詩, which has been translated into English as either "poetry on wandering immortals" or "poetry on wandering into transcendence." Established in the early medieval ages, this kind of poetry features imaginary journeys through the Daoist heavens. The basic themes of the genre have been summarized by David Knechtges as follows: "The *youxian* poem usually portrays the persona traveling far off into the heavens where he visits the haunts of the immortals. He drinks magic potions and eats special immortality-conferring drugs. Some of the pieces contain pure description of immortals and their activities, which the poet recounts rather objectively, with no participation in the traveling and merriment that he describes. More often, however, the poet relates his own imagined flight away from the ordinary world and into the realm of the supernatural."[13]

The first two couplets of "Dreaming Heaven" indeed recall these themes found in poetry on wandering immortals and prepare the key elements for a successful poem of this genre. The first three lines create a chilly, isolated, yet gradually animating picture akin to mental travel. Unlike the stars, which appear to the naked eye to be fixed, the moon moves through the night sky. For the speaker, and some mythical animals living on it, the moon provides both a vantage point for observing the world below and a vehicle of travel through the heavens, moving like a jade wheel that grinds over the dew. Meanwhile, the inhabitants of the moon are introduced as the old hare and shivering toad. They shed sky-blue tears that in the poet's imagination testify to these lonely animals' sadness at their separation from the earthly world. The fourth line then breaks open into a threshold, a cassia-scented path. The sumburgh-bells and the girdle-gems lead the reader to expect the appearance of a goddess, or even possibly an intimate encounter between her and the dreaming speaker.

In the third and fourth couplets, however, Li makes a dramatic turn away from the traditional fantasy of a human entering an immortal wonderland, and veers instead toward a dream-like yet perspicuous vision of the earth. The goddess does not herself appear, while her nightly gaze down on the world below is merged with that of the poet. These two couplets are the much-celebrated climax of the poem, in which ebullient images flash in a grand gesture of the encompassment of both time and space.

The fifth line displays two images that are both natural and emblematic, "yellow dust and clear water," standing for lands and seas, respectively, below the three mountains (*sanshan* 三山), namely the three mythical, everlasting mythical mountains of Penglai 蓬萊, Fangzhang 方丈, and Yingzhou 瀛洲. The sixth line, "A thousand years changed like galloping horses," creates a sense of animation, the unfolding of a number of static images in quick successive movements. Kuo-Ching Tu suggests that the metaphor of "galloping horses" in the line may refer to an ancient toy, the "galloping horse lantern" (*zouma deng* 走馬燈), which he describes as "a lantern adorned with a circle of paper horses which revolves when stimulated by the heat of the light. Revolving endlessly, it is a symbol of the vicissitudes of life."[14] That is indeed the effect that we experience here. In the poem, this effect is created by the tension between the sense of endless historical change and the compressed images of the landscape as the eternal background. The last couplet continues to build upon a dramatic contraction of space. The central kingdom, with its nine divine regions delineated by Yu the Great, shrinks into "nine dots of mist," while all the waves of the seas fill only one cup.

In summary, "Dreaming Heaven" is more equivocal and sophisticated than the rather straightforward celebration of Daoist transcendence typical of poetry on the wandering immortals. As Knechtges points out, such celebration can in general be interpreted in two ways. The first is that "the poet actually believed in the existence" of the transcendental world; the second is that the poet "uses the immortality quest to represent his desire to avoid worldly entanglements."[15] Li He's poem conforms to neither of these interpretations. Although the first half of the poem portrays an immortal realm, the poem shows no indication of a quest for immortality. Nor does it refer explicitly to any circumstances, either personal or historical, from which the speaker wishes to escape. Instead, the poem as a whole creates an animated world image fraught with tension between two poles. One is a nebulous atmosphere represented by the wild dream that frames the entire work and enables mental travel. The other pole, in contrast, is a strong visual sharpness suggesting an attentive eye for dynamism and sophistication, which defies a misty, dreamlike mentality.

To explain fully the effects created in the poem, I propose to situate it in the context of the mid-Tang map culture. I will argue that the unique features of the poem—the mental travel in the celestial realm, the imposition of a historical dimension on the world image, and the deliberate contraction of space—all correspond to the experience of viewing a grand map of the mid-Tang, such as Jia Dan's "Map of Chinese and Foreign Lands." I will also argue that the closing couplet of Li's poem bears a surprising resemblance to that of Du Fu's map reading poem, in that they both juxtapose a misty aura with the image of a cup, creating a strong tension between transcendence and reality. Finally, I will relate the poem to the general views about maps held by the author and his contemporaries, so as to further explore the political vision behind the cartographic eye embedded in the poem.

Jia's map entertains a rich spatial and temporal depth that in many ways resonates with Li's poem. The spatial center of the map is none other than the Chinese territory. The map was based on a Sino-centric world model first introduced in the ancient classic "Tribute of Yu." As Jia writes in his memorial to Emperor Dezong, "The central Chinese territory encompasses five zones of submission and nine provinces; groups with different customs include seven *rong* peoples and six *di* peoples. All under heaven are subjects of the king" (*Zhongxia ze wufu jiuzhou, shusu ze qirong liudi, putian zhixia, mofei wangchen* 中夏則五服九州，殊俗則七戎六狄，普天之下，莫非王臣).[16]

Within the realm of the Chinese territory, Jia imposes an ancient cosmic view to connect the land to the heavens. He uses the method of celestial division (*fenye* 分野), a system that correlates geographical features on the earth with celestial bodies in the sky. For example, it relates the Nine Provinces to the seven stars of the Big Dipper and draws correspondences between the twenty-eight celestial lodges and terrestrial polities. According to the historian David Pankenier, texts dated between the fourth and second centuries BCE alluded extensively to the system, suggesting both a theory of astral-terrestrial correspondence and the idea of mutual resonance between the two realms. Pankenier also argues that unlike the kind of unidirectional influence from above to below found in Ancient Greek works, the Chinese celestial division emphasizes a reciprocal relationship of "as here below, so above."[17] The workings of the celestial system therefore bore direct implications for the security of the earthly states. Examples of this astrologically informed geographical perspective abound in geographic and literary writings of the Tang.[18] What makes Jia's map unique is its success in representing both the celestial and earthly parts of the system on a single map surface.

On Jia's map, because the celestial division system is only applied to the Chinese regions, it creates a strong visual effect that distinguishes Chinese land from non-Chinese land. This effect is well illustrated in the Late Tang poet Cao Song's 曹松 (?–903) poem "Examining the 'Map of Chinese and Foreign Lands'" ("Guan 'Huayi tu'" 觀華夷圖):

> The brush strokes surpass the magic of shrinking the earth.[19]
> I open this map in a moment of peace.
> China belongs to the superior divisions,
> And what constellations match the distant borderlands!
> In a tenth of an inch we distinguish the five sacred peaks.
> By pint and peck we regard the four seas.
> I have always wondered about places unseen.
> Now it is as though I have visited them all.

> 落筆勝縮地
> 展圖當晏寧
> 中華屬貴分
> 遠裔占何星
> 分寸辨諸岳
> 斗升觀四溟
> 長疑未到處
> 一一似曾經[20]

As Der Weerdt comments, in this poem about the author's experience of map-reading, "the poetic persona is cast as a reader who distinguishes between the prominence of the Chinese territories on the map and the marginality of the non-Chinese by rhetorically denying the latter a fixed place in the cosmos. Whereas each Chinese jurisdiction was traditionally correlated with a constellation, the poet is here implying that border areas occupied by non-Chinese populations could not be fixed permanently on the map in this way. The fixed coordinates lock the Chinese territories in a permanent cosmic order against historical change, while leaving the non-Chinese areas outside this system."[21]

This fixed system of virtually permanent geographical formations in Chinese land and their celestial correspondents also appeared in another mid-Tang grand map, Li Gai's "Map of the Territorial Record 地志圖." Lü Wen, an important mid-Tang intellectual, describes this system in his preface to Li Gai's map in the following way: "Every [earthly] boundary, represented in [the map measuring tiny] square inches, corresponds to celestial divisions in the world above. The [correlation between] celestial phenomenon and earthly features can thus be observed with splendid clarity" (*Fangcun zhijie, er shangdanghu fenye. Qianxiang kunshi, bingyan keguan* 方寸之界，而上當乎分野。乾象坤勢，炳焉可觀).[22] That is to say, this representational system inspires in viewers a sense of cosmic perpetuity as they witness the unity of earth with heaven, while it simultaneously reinforces their perception of the boundaries on earth.

The ahistorical effects of celestial division notwithstanding, Jia manages to make the presence of history conspicuous by imposing a contrasting cartographic scheme on the map. He uses red ink for present-day place names and black ink for historical names adopted and abandoned by dynasties of the past. As Jia reports to Emperor Dezong in his memorial, "ancient commandery and fiefdoms are marked in black; present-day prefectures and counties are marked in red" (*Gujunguo tiyi mo, jinzhouxian tiyi zhu* 古郡國題以墨，今州縣題以朱).[23] This approach of graphical organization at once ties the present place to and differentiates it from its history. It adds another layer of textual complexity to the map and offers its readers a strong yet nuanced sense of shared cultural tradition. Its visual effects lend the map temporal depth to complement and complete the three spatial dimensions it represents.

The grand maps could easily evoke a sense of mental travel through time and space among their viewers. As Jia describes his map to Emperor

Dezong: "Despite containing a vast cosmos, the opened map occupies less space than a courtyard; every place boats and carriages can reach, the map presents to the viewing eye" (*Yuzhou sui guang, shuzhi bu yingting; zhouche suotong, lanzhi xian zaimu* 宇宙雖廣，舒之不盈庭；舟車所通，覽之咸在目).[24] For him, the "Map of Chinese and Foreign Lands" contracts the cosmos and encourages a viewer's eye and mind to travel through it. Lü Wen similarly describes the compelling effects of viewing the "Map of the Territorial Record":

> All under heaven is contained on a wall in the room. Within the doors the four seas are encompassed; inside the windows the eight corners of the world are encapsulated. Glorious mountains and great rivers, one can reach at will; foreign lands and remote realms, one can survey at whim. All lands of the Chinese can be seen from high above; the past and present can be traversed as one sits unmoving. Beholding all the territories of emperors and kings, one appreciates the magnitude of the cosmos.
>
> 普天之下，盡在屋壁。戶納四海，窗籠八極。名山大川，隨顧奔走。殊方絕域，舉意而到。高視華裔，坐横古今。觀帝王之疆理，見宇宙之寥廓。[25]

This is a telling statement from a map viewer's perspective. Within a single map, he can simultaneously see the natural landscape of his world and the unrelenting trends of history. He visualizes the boundaries dividing the Chinese and the non-Chinese, and the vastness of the cosmos in which both reside. This awe-inspiring cartographic presence, along with its holistic representational method of celestial division, elucidates the sense of grandeur that both Jia and Lü evoke: it was not merely a rhetorical gesture, but rather reflected the intense map viewing experience shared by mapmakers and map readers alike.

Rereading Li's poem "Dreaming Heaven" in light of Jia's grand map, we cannot help but wonder at the resemblance between the images created by the second half of the poem and the view of the world as represented on a grand map. The coexistence of celestial division and human history on Jia's map mixes constants and variables, generating a fabric of space-time whose visual complexity can elicit in the viewer a sense of seeing history mutate against the backdrop of an everlasting cosmos. This effect is consubstantial with the tension in the third couplet of Li's poem between the eternal terrestrial features and an ever-changing history that flows "like galloping horses." In the fourth couplet, the line "Gazing afar toward the middle land that appears like nine dots of mist" sustains the perspective from a remote distance but shifts the focus from the supernatural mountains to the Chinese land, as if to highlight its central position on a map of the world. Subsequently, "All the seawater filling a cup" continues to contract the seas from the periphery, thus completing the poem with the full image of a China-centered world as viewed from the heavens.

Moreover, if we compare the last couplets of Du Fu's poem on the painting-map with Li's poem, we see that they in fact follow a similar pattern. Both speakers start with a misty image and then use the image of a cup as an anchor. Let us reread the closing passage of Du Fu's poem:

> My excitement joined the glowing mists and clouds,
> How fortunate that my pure cup was never empty.

And then compare it with that written by Li He:

> Gazing afar toward the middle land that appears
>   like nine dots of mist,
> All the seawater filling a cup.

In Du's couplet, "the glowing mists and clouds" hint at transcendence, while the filling of the wine cup evokes a sense of social reality. Similarly, in Li's couplet, the "nine dots of mist" create a hushed transcendental atmosphere before the abrupt action of filling the cup with sea water. Without Du's poem as a reference point, the cup in the last line of Li's poem is simply a metaphorical device intended to highlight a contracted image of the sea. We find a similar expression in Cao Song's poem about viewing Jia's grand map: "By pint and peck we regard the four seas." But Du's poem inspires us to understand Li's cup also as an actual wine cup, presumably in the poet's hand, which brings the speaker's mental journey back into a setting of ordinary human life. Between the misty images of transcendence and the cup, both poets complete a circle from an expansive scope to the image of the cup, which often serves as a symbol of the poet's passion and enjoyment.

Certainly, these two poems address different contexts. Du's is explicitly about map viewing situated in the setting of a daytime banquet, while Li's couplet suggests mental travel at night and does not actualize its visual references and surrounding space. Yet the poems are united in their fresh ways of creating spatial imaginaries. Looking at an actual map, Du sees beyond the topographies of Sichuan to a wider imperial space. Without mentioning a map, Li condenses and internalizes the perspective, patterns, and dynamics exemplified by a grand map.

Furthermore, the grand vision of the Chinese landscape in Li He's poem has a political dimension that resonates with a mid-Tang map culture that was marked by a general preference for grand map over maps of a more limited scope. For the literati of Li He's generation, the image of the all-encompassing empire represented on a grand map was intrinsically associated with a grand political vision of imperial restoration. For instance, in his preface to a historical map called "Territorial Map of the Han Dynasty" ("Han yudi tu" 漢輿地圖), Lü Wen reflects on how the founder of the Eastern Han, Emperor Guangwu (5 B.C.E –57), used territorial maps as a blueprint for retaking lost lands, and how from the very beginning Emperor Guangwu already had the whole scope of the empire in mind. As Lü puts it,

> As for Emperor Guangwu's ambition to restore imperial power, he had only to open the territorial map and a scope of ten thousand *li* sprang to his mind. Would he acquire a county and only then think of acquiring a prefecture, acquire a prefecture and only then think of acquiring a bigger district? So great was Guangwu's

> ambition! This is how he managed to continue the offering of sacrifices to the Han ancestors, to match heaven's will, and to preserve the old traditions.
>
> 若夫光武恢復之志，則一披輿圖，而萬里之幅員，皆入於靈府。豈嘗得一邑而始思得一州，得一州而始思得一部？大矣光武之志也！斯其所以祀漢配天，不失舊物歟。[26]

What Lü suggests learning from the Han case is to think in terms of the big picture. Success on Emperor Guangwu's level was built on a mindset of a range comparable to an imperial map; finite vision and provincial aspiration would never have sustained such a definitive revival of the Han Empire.

Although Li He left no works that explicitly address a grand map, he, like Lü Wen, was deeply concerned with the imperial recovery of previously held territory. As he famously proclaims in a different poem: "Why shouldn't a young man wear a *wu* sword? / He could win back fifty prefectures in the passes and mountains" (*Nan'er hebu dai wugou, shouqu guanshan wushizhou* 男兒何不帶吳鉤，收取關山五十州).[27] The "fifty prefectures" are the imperial territory of the Hebei region lost to powerful warlords who had plagued the empire for decades. In a long poem titled "On Leaving the City and Parting from Zhang Youxin I Pledge to Li Han with Wine" ("Chucheng bie Zhang Youxin chou Li Han" 出城別張又新酬李漢), Li He describes the grand imperial enterprise and the prosperity of the capital in the following couplet: "The imperial picture [also: map/governance] spans the four seas, / And all our citizens are gentlemen" (*Huangtu kua sihai, baixing shi changshen* 皇圖跨四海, 百姓施長紳).[28] The term *huangtu* 皇圖 was originally a synonym for *hetu* 河圖, which refers to the ancient secret diagram used to legitimate the rule of ancient sage kings.[29] In other words, contained in the original *huangtu* is an oracle, not a map. But the word *tu* and the spatial connotation in the expression "spans the four seas" indicate that the imperial icon of *huangtu* and the all-encompassing grand map are ideologically synonymous. Both are magnificent icons and powerful expressions of the empire's expanse and restoration.

Like Lü Wen, Li He distains a limited regional view. His poem "Song of the Young Five-Grain Pine with Preface" ("Wuli xiaosong ge bing xu" 五粒小松歌并序) offers a case in which the author associates a prefectural map with an inferior personality:

> Cultivated Talent Xie and Du Yunqing once asked me to write a song for a young five-grain pine. I was very busy compiling materials for my books at the time and could not compose lyrics for them. Ten days later I jotted down these eight lines to fulfill their request.
>
> Snake's son, snake's grandson,
> Scales coiled like a dragon's.
> My grains, new and fragrant,
> Were food for Hong Ya.
> Leaves lapped in green wavelets,
> Glossy and rich.
> Neat bundles of dragons' whiskers,
> Trimmed off with scissors.
> On my owner's wall
> Maps of the prefecture.
> Round my owner's hall
> Mobs of stodgy scholars.
> Bright moonlight, white dew,
> Autumn tears falling.
> Pointed stones, stream clouds,
> May I send you this letter?[30]
>
> 前謝秀才杜雲卿命予作五粒小松歌，予以選書多事，不治曲辭，經十日，聊道八句，以當命意。
>
> 蛇子蛇孫鱗蜿蜿
> 新香幾粒洪崖飯
> 綠波浸葉滿濃光
> 細束龍髯鉸刀翦
> 主人壁上鋪州圖

主人堂前多俗儒
月明白露秋淚滴
石筍溪雲肯寄書[31]

The poem was written when Li was serving as a low-ranking official in Chang'an. It is often read as an autobiographical work that expresses the poet's frustration as an unappreciated talent.[32] According to this reading, the poet here assumes the persona of the five-grain pine (*pinus pentaphylla*), which was regarded by Daoists as a lofty species of supernatural origin whose seeds aided longevity. As the first several lines depict, the tree's bark resembles the scales of snakes and dragons, and its grains are offerings to the Daoist immortal Hong Ya. Sadly, this unusual tree is planted in the courtyard of an undeserving host who has a "map of the prefecture" on his wall and many "stodgy scholars" in his hall. The tree cries and hopes to send this letter to its old companions, the "pointed stones" and the "stream clouds," which surrounded it when it was growing in the natural environment. In the poem, Li creates a parallel and a rhyme between the prefectural map (*zhoutu* 州圖) and the "stodgy scholars" (*suru* 俗儒). In this context, a regional map is harshly dismissed and thoroughly caricatured because of its limited scope. For a poet of grand vision, maps of limited scope are intrinsically inferior: they fail to remind viewers of the heights the empire had once reached; their provincial radius prevents aspiring warriors from seeing through the fog of total war; and as the Qing commentator Wang Qi points out, the quality of the painting itself is usually inferior.[33] Such maps could only serve to embody the parochial mindset that hindered an imperial renaissance.

To conclude, Li was a poet of the big picture, both in imaginative and political terms. His vision in "Dreaming Heaven" was elevated to the level of the heavens from which to look down upon the world, which can be seen as an aestheticization of his political passion. In this case, introducing map culture helps us better understand the poem from a productive interpretive perspective. By the time Li endeavored to create new poetic visions of the cosmos and the empire, mapmakers like Jia

Dan had already operated on the same kind of high level of visual representation. Considered separately, poetry and maps demonstrated intriguing resemblances. Together, they formed a synergy even more illuminating than their powers employed on their own. In this sense, the grand map emerges as a metaphorical carrier for the poet's artistic and social pursuits. The poem, in turn, brings out the full potential of the cosmic imagination that a mid-Tang grand map could sustain.

GRAND MAPS AND IMPERIAL SPACE IN LIU ZONGYUAN'S WRITING OF THE PERIPHERIES

If the conversation between grand maps and literature in Li He's case is manifested in his all-encompassing cosmic vision, then in the case of Liu Zongyuan, this conversation is mostly found in his writings about the imperial peripheries. When Liu was demoted to the southern hinterlands of the empire, he developed new geopolitical perspectives and literary techniques thanks to his knowledge of maps, especially grand maps. Such perspectives and techniques inspired him to creatively manipulate distances and scales through literary representation, and allowed him to delineate a comprehensive and carefully calibrated view of the empire. From this new literary vantage point, he attained a sense of authority to negotiate with the imperial center as a marginalized politician and intellectual voice.

In 819, when Liu served as the prefect of the peripheral Liuzhou 柳州 (present-day Liuzhou, Guangxi Province), the imperial court ordered Liu to submit a special regional map under the title "Map of Faraway Lands," in addition to collecting and reporting on local customs and histories. Instead of complying immediately with the central government, Liu offered a poem as a response titled "The Southern Department Sent an Official Letter to Request the Creation of a 'Map of Faraway Lands' and a Comprehensive Study of the Local Customs and Histories" ("Nansheng zhuandie yuju 'Jiangguo tu' ling jintong fengsu gushi" 南省轉牒欲具江

國圖令盡通風俗故事), in which he invokes Jia Dan's "Map of Chinese and Foreign Lands":

> In our sacred dynasty the imperial territory
> has extended to the seashores.
> Even distant desolate areas deserve a record
> of their mountains and rivers.
> The first map to include this was the "Map of Chinese
> and Foreign Lands."
> Even the *Record of Local Customs* did not have it.[34]
> It is hard to ask questions of the old people with hair
> pulled up in the shape of a hammer,
> Who dares to linger in the deep realm of the Miao?[35]
> If the South Palace intends to search for remaining customs,
> Try consulting "The Meeting with the King"
> in the *Zhou Documents.*[36]

> 聖代提封盡海壖
> 狼荒猶得紀山川
> 華夷圖上應初錄
> 風土記中殊未傳
> 椎髻老人難借問
> 黃茆深峒敢留連
> 南宮有意求遺俗
> 試檢周書王會篇[37]

This is one of LiuZongyuan's last poems, since he died that same year. In the title, the "Southern Department" (*Nansheng* 南省) refers to the Department of State Affairs (*Shangshu sheng* 尚書省), which was located to the south of the Palace of Great Light (Daming gong 大明宮) in Chang'an.[38] The term *jiangguo* 江國 originally referred to a small state of the Spring and Autumn period, and is here borrowed to refer to the faraway land of Liuzhou.[39] The center of power as represented by the Department of State Affairs and the imperial periphery is connected by the imperative predicate of "Sent an Official Letter to Request." With this, a hierarchical spatial structure of the empire is immediately established.

The main body of the poem succinctly captures the dilemma faced by the Tang government on the peripheries. On the one hand, the government had shown impressive achievements in recognizing and representing faraway lands. The first two couplets celebrate the progress Jia Dan's grand map has made; the third line remarks that the map was the first in history to incorporate Liuzhou, where the poet's current post sits. The sense of amazement that the character *chu* 初 (for the first time) carries implies the new possibilities for cartographic exploration that the map has opened up. However, the celebration of imperial expansion comes to a sudden halt here. The speaker holds back and, thinking in his role as a frontier official, starts to question the government's ambition to exercise direct control over the local land. The first critique he raises is practical: it is virtually impossible for him to go deep into the frontier and extract information about local customs. How is he going to even communicate with the aboriginals, those "old people with hair pulled up in the shape of a hammer"? The language barrier and cultural differences would certainly sabotage the imperial project. The second critique, which is more political, is that the imperial government, here again represented by the Department of State Affairs, or the "South Palace" 南宮, may risk deviating from traditional ways of dealing with ethnic minorities. Advising tacitly against imperial overreach, Liu recommends that the court follow the example of the ancient Zhou kings and encourage submission among faraway peoples through the demonstration of moral rectitude.

This poem contains Liu Zongyuan's only explicit reference to Jia Dan's map. The circumstances of his first viewing of the map are unknown, but his familiarity with the cartographic perspective exemplified by grand maps can be seen in others of his works that masterfully engage with the relationship between center and periphery. The first such work we will examine is titled "A Poem Sent from Liu Zhou to Relatives and Old Friends in the Capital" 柳州寄京中親故:

Linyi's mountains spread into the autumn of the malaria sea.

Zangke River flows in front of the prefectural city.
It was kind of you to ask me where Dragon City is.
Traveling due north for three thousand *li* one would reach Jinzhou.

林邑山聯瘴海秋
牂牁水向郡前流
勞君遠問龍城地
正北三千到錦州[40]

In the first couplet, Liu Zongyuan establishes the geography of Liuzhou, also called Dragon City, as is shown in the third line. Linyi is in modern Vietnam, and the Zangke River flows through Guangxi into the South China Sea.[41] Both are place names that would evoke a sense of extreme remoteness for people in the capital. Moreover, the character "malaria" (*zhang* 瘴) in the first line is typically associated with an alien, dangerous southern atmosphere in Tang literature. Accordingly, the pairing of "malaria" 瘴 with the "sea" (*hai* 海) further adds to the couplet a sense of overwhelming harshness.

In the second couplet, Liu makes a dramatic twist to impress his implied readers in the capital. Instead of spelling out the actual distance between Liuzhou and Chang'an, he presents a carefully chosen reference point, Jinzhou (close to modern Mayang County in Hunan Province), to initiate the reader's own spatial reasoning. This is a striking way to play with space and distance, because Liu appears to be manipulating the predominant, imperially sanctioned cartographic system of the time. His indication of direction in "due north" (*zhengbei* 正北) followed by the distance of "three thousand *li*" (*sanqian* 三千) is not meant to be literal. Rather, it is a term specifically used in the coordinate system of "Four Extremes and Eight Extents," a system fully established in the mid-Tang by Li Jifu's *Maps and Treaties of the Prefects and Counties of the Yuanhe Reign*.

According to the science historian Wang Qianjin, this system created an extensive database of measurements of distance, which was instrumental in the creation of mid-Tang grand maps.[42] The system represents the locations of administrative units and the distances between them using their relative positions in the empire. It sets Chang'an and Luoyang, the two capitals, as primary reference points from which to measure the relative positions and distances of prefectural seats. It then further specifies the location of these prefectural seats by way of their distance from other nearby prefectural seats. The system also replicates this methodology at the prefectural and county levels. Although named "four extremes and eight extents," the system can pinpoint any position with up to sixteen directional references: four cardinal directions of due north (*zhengbei* 正北), due south (*zhengnan* 正南), due east (*zhengdong* 正東), and due west (*zhengxi* 正西), four intercardinal directions that combine the four cardinal directions above, and eight secondary intercardinal directions, such as east by northeast (*zhengdong weibei* 正東微北).[43] A standard mid-Tang description of Liuzhou's relative position in the system would be something like what Li Jifu offers in in *Maps and Treaties of the Prefects and Counties of the Yuanhe Reign*:

> Eight extents: 4245 *li* due north to the Upper Capital, 4140 *li* northeast to the East Capital, 540 *li* east to Guizhou, 160 *li* east to Xiangzhou, 223 *li* by land due north to Rongzhou, 380 *li* by water.
>
> 八到：北至上都四千二百四十五里。東北至東都四千一百四十里。東至桂州五百四十里。東至象州一百六十里。北至融州陸路二百二十三里，水路三百八十里。[44]

In Liu Zongyuan's representation of Liuzhou's relative location in the empire, in contrast, we are told that "traveling due north for three thousand *li* one would reach Jinzhou." In this unusual way, he creatively deploys the standard cartographic terminology from "Four Extremes and Eight Extents," as much for poetic aesthetics as for spatial-political expressivity. As is clear in the description earlier from *Maps and Treaties of the Prefects and Counties of the Yuanhe Reign*, the standard positional reference point in the northern direction for a southern city like Liuzhou is Chang'an. Liu could therefore have offered a much more straightforward description of Liuzhou's location as four thousand *li* south of the capital city. Yet, as an ostracized official who actually traveled from Chang'an to Liuzhou, he defies that set practice. He references Jinzhou, already a hinterland in the middle of nowhere, as an unconventional midway point. The line marks distance as if in a relay across the empire and imparts a stronger sense of isolation to the image of Liuzhou. Moreover, to travel from Liuzhou to Jinzhou, one only needs to follow a land route no more than one thousand *li*, not three thousand *li.*[45] This would not have been an obscure fact to Liu, as Liu had served in Yongzhou (also in Hunan) for ten years, and had traveled from Chang'an to Liuzhou by way of Hunan. Liu the poet nonetheless chooses to exaggerate the distance to a staggering three thousand *li*. The representational effects of the term "three thousand *li*" dramatize the isolation of Liuzhou as the standard but plain description of "one thousand *li*" would not.

The poem therefore locates peripheral landscapes on an imaginary imperial map. While the first couplet is confined to an artistic organization of place names on a regional scale, the second couplet expands the scope of imagination to the entire imperial realm. Liuzhou, Jinzhou and Chang'an belong to three distinctive regions of the empire that are distant one from another. The visualization of a line that connects the three places can thus only be premised upon the kind of large-scale spatial imagination offered by a grand imperial map. This grand view is further mediated by the poet's creative employment of the cartographic vocabulary from the "Four Extremes and Eight Extents" system. The poem is a telling testament to his indebtedness to contemporary cartographic innovation, as well as his poetic freedom from it. As such, the poem tells its audience in the capital several things. First, it offers a geographical portrayal of the exotic landscape of Liuzhou. Second, it reconnects Liu Zongyuan to his old elite circle in the capital from which he was ostracized. Most importantly, through a deliberate manipulation of the standard geographic knowledge in the poetic space, the poet articulates a strong protest against the authorities who had banished him to this bleak place.

Another example of Liu Zongyuan's creative negotiation with the imperial space through a map-informed spatial manipulation is his landscape essay "Record of an Excursion to Huang Creek." The essay was written in 813, after Liu had been banished from the capital to Yongzhou 永州 (present-day Yongzhou city in Hunan Province) for eight years, and it records Liu's visit to a creek close to the eastern border of the Yongzhou area. Here is the essay in its entirety:

> Of the hundreds of prefectures known for their mountains and waters in the area northward to Jin, westward to Bin, eastward to Wu, and southward to the area between Chu and Yue, Yong is the best. Of the hundreds of villages known for their landscapes in the prefecture covering a hundred *li* to the Wu Creek in the north, Xiangyuan County in the west, Shuang Spring in the south, and East Village on Huang Creek in the east, Huang Creek is the best.
>
> Huang Creek is seven *li* from the capital of the prefecture. Walking from Eastern Village southward for six hundred paces, I reached the Temple of Lord Huang. Above the temple, mountains tower like walls on both sides. Flowers like red cinnabar and leaves like green jade grow in rows, extending up and down with the mountain range, interrupted only by cliffs and grottoes. Small rocks spread out evenly in the water. Pulling up my robe, I walked from above the temple eighty paces to where the water reached the First Pool. It was wondrously beautiful and could hardly be described. In its general contours, it looked like a huge urn slashed open right in the middle, with the surrounding cliffs standing up a thousand feet above the ground. The water of the creek gathered here like eyebrow pigment and thick oil. The creek flowed into the pool like a white rainbow and yet made no sound at all. Hundreds of fish came and gathered under the cliffs. Going further south for a hundred paces, I reached the Second Pool. Here, facing the torrent of the creek were towering cliffs in the shape of chins and roots of teeth. Below, huge rocks lined up randomly, upon which one could sit down to eat and drink. There was a bird with a red head and black wings, as big as a swan, standing with its face to the east. From here, further southward for several li, the landscape appeared all the same, with the trees more imposing, the cliffs

sharper, and the flowing water resounding all the time. One *li* further south, I reached a grand river. Here the mountain spread out and the water slowed, and there were fields under cultivation. When Lord Huang was still a human, he lived here.

According to legend, Lord Huang was of the surname Wang and a descendant of Wang Mang. After Wang Mang died, he changed his name to Huang and escaped here to choose a hidden and cragged spot to hide. Earlier, Wang Mang had said: "I am the descendant of Huang and Yu (i.e., the Yellow Emperor and the sage-king Shun)." Therefore he gave his daughter the title "Princess of the August House of Huang." "Huang" and "Wang" sounded very much alike; since there was such historical evidence, the legend became all the more credible. Since Lord Huang resided here, the people were able to live in peace. They all thought he had attained the Way; therefore, after he died, they worshipped him and set up a shrine in his honor. Later the shrine was moved closer to the people. Nowadays it is by the creek on the north side of the mountain. On the sixteenth day of the fifth month of the eighth year of the Yuanhe reign, after I returned from my trip, I wrote the above record to enlighten future visitors.[46]

北之晉，西適豳，東極吳，南至楚越之交，其間名山水而州者以百數，永最善。環永之治百里，北至於浯溪，西至於湘之源，南至於瀧泉，東至於黃溪東屯，其間名山水而村者以百數，黃溪最善。

黃溪距州治七十里，由東屯南行六百步，至黃神祠。祠之上兩山牆立，丹碧之華葉駢植，與山升降。其缺者為崖峭巖窟，水之中皆小石平布。黃神之上，揭水八十步，至初潭，最奇麗，殆不可狀。其略若剖大甕，側立千尺，溪水即焉。黛蓄膏渟，來若白虹，沉沉無聲，有魚數百尾，方來會石下。南去又行百步，至第二潭。石皆巍然，臨浚流，若頦頷斷齶。其下大石雜列，可坐飲食。有鳥赤首烏翼，大如鵠，方東嚮立。自是又南數里，地皆一狀，樹益壯，石益瘦，水鳴皆鏘然。又南一里，至大冥之川，山舒水緩，有土田。始，黃神為人時，居其地。

> 傳者曰："黃神王姓，莽之世也。莽既死，神更號黃氏，逃來，擇其深峭者潛焉。"始莽嘗曰："余，黃、虞之後也。"故號其女曰黃皇室主。黃與王聲相邇，而又有本，其所以傳言者益驗。神既居是，民咸安焉。以為有道，死乃俎豆之，為立祠。後稍徙近乎民，今祠在山陰溪水上。元和八年五月十六日，既歸，爲記，以啓後之好游者。[47]

The essay consists of three sections. The first singles out Huang Creek as the best scenic spot within a vast geographical space. The second is a vivid description of the landscape along the creek, from the shrine to Lord Huang through the richly described waters, stones, and animals, to an open area of human habitation. The third lends historical depth to Lord Huang's divine image with a recounting of his eventful life from a dubious figure associated with Wang Mang 王莽 (45 BC–23 AD), the notorious usurper of the Western Han Dynasty, to a virtuous hermit in Yongzhou.

In this section, I will argue that new insights into the content and form of this essay can be gained by reading it in relation to Jia's grand map. A cartographic perspective enabled Liu Zongyuan to operate his narrative on an imperial scale while still depicting a peripheral landscape and expressing moral sentiments. The first part mimics the experience of reading a grand map, and the second part follows patterns similar to that of the textual explanations of landscape that accompany a grand map. Taken together, the two parts prepare the ground for the third part about the local figure Lord Huang. The essence of Lord Huang's story has to do with the possibility that an exile may redeem his past and be remembered as an honorable figure in history. This thinly veiled historical lesson for Liu as a demoted official, however, is not a local story alone, but a story that connects the periphery of the empire to its center. In this sense, the essay develops both consistently and coherently with Liu's understanding of the imperial space informed by grand maps.

The first part of the essay has drawn interest from many critics. To date, most critics have shared the assumption that it is centered on a rhetorical gambit not unusual in the history of Chinese prose. Traditional commentators from Wang Yinglin 王應麟 (1223–1296) to Lin Shu 林紓 (1852–1924) tend to read it as imitative of Sima Qian's (145– or 135–90 BC) "Biographies of Southwestern Barbarians" 西南夷列傳, which also starts with a progressive comparison according to the author's spatial orientation. They praise Liu for his ingenious appropriation of the classical historical style for landscape writing.[48] He Zhuo 何焯 (1661–1722), who also takes Liu's imitation for granted, believes that this part of the poem is derivative and not necessarily true to life, and therefore the essay might do better without it.[49] The modern literary historian Xiaoshan Yang rebukes He's suggestion that this be omitted by pointing out that the Liu's appropriation of the historical tone is thematically relevant to the essay's implied moral lesson. At the same time, Yang points out that Liu's opening strategy was not unusual in the Tang because landscape writings such as Bai Juyi's "Record of the Cold Spring Pavilion" ("Lengquan ting ji" 冷泉亭記) shared a similar rhetorical pattern, in which authors name "bests" in successive groups of similar objects in different geographical spaces.[50]

Taking into account the development of grand maps in the mid-Tang, we can read the first part of Liu Zongyuan's essay from a new angle. The writing mimics the zooming-in effect of map reading and immediately pinpoints the local site within a broad imperial time-space. In the first sentence, five place names are invoked: Jin, Bin, Wu, Chu, and Yue. They denote five ancient states and prefectures. In four directions, the five places collectively embrace a vast territory, spanning nearly all of the Chinese empire.[51] In effect, the first sentence evokes an aerial view of China from high above, and, as if guiding the reader to scan a large-scale map, the narrator quickly zooms in on Yongzhou as his locus of interest. The second sentence sees a dramatic diminishing of the visual scope, from the sweeping imperial space to an area of "a hundred *li*." The narrator's selection process then continues, finally reaching the Huang Creek as the most beautiful place in the world.

Read from a cartographic perspective, the first part is decisively different from other passages that previous critics have found to be similar. For instance, Sima Qian's "Biographies of Southwestern Barbarians" opens in the following way:

> There are dozens of chiefs ruling among the southwestern barbarians, but the most important is the ruler of Yelang. To the west of Yelang live the chiefs of the Mimo, of which the most important is the ruler of Dian. North of Dian live numerous other chiefs, the most important being the ruler of Qiongdu. [52]
>
> 西南夷君長以什數，夜郎最大；其西靡莫之屬以什數，滇最大；自滇以北君長以什數，邛都最大。[53]

In the passage, Sima Qian first names Yelang as the most powerful chieftain among all the barbarian chiefs in the southwest, and then circumscribes the scope to the peoples west of the Yelang tribe to highlight the chief of Dian. Toward the end, he moves his focus to the chiefs north of Dian, among whom he singles out Qiongdu as the most important. Like Liu Zongyuan's essay, this passage conveys a clear sense of relational spatial configuration and then names important subjects in each region under examination. But the overall scope of the passage does not exceed a relatively small, homogenous region. More importantly, Sima Qian organizes his narrative as a series of departures from one region to another: from Yelang's territory to the land to his west, and then from Dian's territory to the area to his north. The narrative does not involve an overarching grand view and is different in spatial organization from the zooming-in structure of Liu's essay.

Another example comparable to Liu's essay is Bai Juyi's "Record of the Cold Spring Pavilion." This is how Bai introduces the Cold Spring Pavilion:

> Among the mountains and rivers in the southeast, those in Yuhang prefecture are the best. Within the prefecture, the Lingyin Temple

> is the most remarkable. Within Lingyin Temple, the Cold Spring Pavilion is the most outstanding.
>
> 東南山水，餘杭郡為最。就郡言，靈隱寺為尤。由寺觀，冷泉亭為甲。[54]

This passage resembles Liu's essay in that it also uses a progressive process to lead readers' focus to the real subject of interest. A closer look at the place names, however, shows that the geographical areas from which Bai selected his "best" subjects are much vaguer and more generalized than those found in Liu's passage. Bai crowns the famous administrative unit Yuhang prefecture as the best in the rather vague, undefined space of the southeast, and then goes directly to a clearly demarcated religious site. For Bai, the progressive strategy is indeed a rhetorical device rather than a serious engagement with the imperial space. It does not involve an implied eye that scans across a number of regions on a broad surface and only then zooms in on one specific place, which is what we find in Liu's essay.

In addition to a spatial scope that mimics a grand map, the first part also involves a historical perspective. The five place names, Jin, Bin, Wu, Chu, and Yue, are all historical names. We have seen that Jia Dan noted both historical names and contemporary names for each place, marking them in different colors. His map therefore has a historical dimension that links the places to a shared cultural tradition. Here, Liu Zongyuan also establishes an imperial time-space with both spatial width and temporal depth. To be sure, at the time, it was very common for a literary writer to use historical names to refer to places, and Liu was but one of many writers who did so. However, when combined with geographical directions and immediately followed by a zooming-in effect, the historical names greatly enhance the resemblance between the first part of Liu's essay and Jia's grand map, in that the passage would evoke a similar feeling from its readers as Jia's grand map would from its viewers.

The second part of the essay is focused on Liu Zongyuan's experience of an excursion along the Huang Creek. The narrator works his way through a series of local spaces, carefully taking note of the delicate topography. If the previous part reminds us of a map-reading eye, then this part bears a similarity to the annotations of specific places attached to Jia's grand map. As discussed in the Introduction, traditional Chinese maps typically come with substantial textual notes, either on the map itself or attached to the map as an appendix, that explain details about the geographic features represented therein. In particular, Jia's grand map was known for its forty volumes of map notes titled "Notes on Administrative Jurisdictions and Foreigners in Four Directions." These notes, a compilation of Jia's work over several decades, themselves constituted a geographical encyclopedia. Unfortunately, only one note in this great work is extant. Nevertheless, this singular evidence of Jia's map notes is helpful when looking at the second part of Liu's essay, which describes Jade Mountain in Jiangxi Province:

> Jade Mountain stands as high as the sky, and gestures toward the Northern Dipper, and is therefore also called Jade Dipper Mountain. Moving up and down along the foot of the mountain, after fifteen *li* one reaches the Dayang Slope, and the land there is wide and open for several hundred acres. Surrounding this are unusual peaks and charming mountain ranges, strange rocks and deep pools. This is truly the grotto-home of immortals and spirits.
>
> The mountain has eighteen waterfalls on the Dragon Pool and twenty-four wonders: Jade Ye Peak, Silver Point Peak, Seven Circle Peak, Lion Peak, Stone Ox Peak, Cloud Canopy Peak, Coiled Dragon Ridge, Gold Rooster Mound, Washing Ink Pond, Anticipating Fragrance Mound, Nine Lotuses Pond, Heaven Gate Peak, Flying Spring Peak, Screen Peak, Shipo Rock, Bathing Buddha Pond, Colorful Cloud Rock, Cloud-Passing Cave, Intertwined Trees, Heavenly Sage Pine Tree, Diamond Ridge, Stone-Drum Peak, Aluohan Peak, and Zhichu Rock. This mountain is truly the best

in the region for its natural wonders, and so the county is named after it.

> 玉山與天際，勢聯北斗，又名玉斗山。循山之麓升降，凡十有五里，至大洋阪，地寬曠約數百畝，而奇峰秀嶺，恠石深池，環列于前後左右，真仙靈之窟宅也。
>
> 山有龍潭一十八磜，又有二十四奇，曰玉瑯峰、銀尖峰、七盤峰、獅子峰、石牛峰、雲蓋峰、蟠龍岡、金雞墩、洗墨池、望香墩、九蓮池、天門峰、飛泉峰、屏風峰、誓坡石、浴佛池、彩霞巖、過雲洞、連理木、天聖松、金剛嶺、石鼓山、羅漢峰、志初巖。真一邑勝概之尤者也，故縣亦由此名。[55]

In this passage, Jia assumes the voice of a literati traveler. After a brief introduction to the name of the mountain, the narrator begins to "move up and down along the foot of the mountain" to discover open land surrounding it. Looking around, the narrator further discovers that the open land is surrounded by geographic wonders like strange rocks and deep pools. He exclaims at the spiritual aura of the place, and then, like a landscape explorer, he begins to list the names of the sites of natural beauty in the region, which includes eighteen waterfalls and twenty-four wonders. At the end of the passage, the narrator lauds the mountain as "the best" in the region.

Admittedly, Liu Zongyuan did not necessarily have this passage in mind when he was writing about the surroundings of the Huang Creek in the second part of his essay. Stylistically, the differences between Jia and Liu's passages are salient. Compared to Jia the geographer, Liu the literary master showed much more acute sensitivity to the visual and aural aspects of the landscape. Nevertheless, some similar patterns between the two suggest that Liu's landscape depiction can be situated in the tradition of geographic writing. For example, while Jia describes the implied traveler moving up and down along the mountain (*xun shanzhilu shengjiang* 循山之麓升降), Liu describes the plants as "extending up and down the mountain range" (*yushan shengjiang* 與山升降). The two expressions entertain the same sense of movement inherent in the landscape. Liu then depicts the natural surroundings as he walks down along the creek, coming across first a pool, and then a second pool, before reaching an open land at the foot of the mountain. In both Jia's map notes and Liu's landscape writing, the contours of the terrain are outlined through a contrast between the unusual objects in a mountainous enclosure and the smooth, soothing open landscape at the foot of the mountain. Finally, from the first part of Liu's essay, we know that the focus on Huang Creek is the result of a process of selection from among hundreds of landscapes in the region. This way of perceiving landscapes in an enumerative manner echoes the second half of Jia's map notes as well.

Of course, with the evidence currently available, it would be hard to ascertain whether Liu Zongyuan wrote his essay under the direct influence of Jia's map notes. Rather, it is more likely that Jia's map notes and Liu's landscape essay share roots in the same historical tradition of geographic writing. As discussed in the Introduction, the conventions of geographic writing were established in the Six Dynasties. With the addition of Jia's grand map and map notes, we can now read the first and second part of Liu's essay as an organic whole. If the first part mimics a map-reading eye, then the second part parallels a map note, helping visualize the specific geographic spot on the map on which the implied eye focuses. Together, they provide a spatial background against which Lord Huang can emerge. In the third part of the essay, Huang's life story as an exile inserts into this geographically inspired landscape writing a layer of historical meaning.

The third part, many critics believe, is the heart of the essay. In this view, Liu finds a kindred spirit, or an image of himself, in the story of Lord Huang. However, that Liu traced the story of Lord Huang back to the notorious historical figure of Wang Mang has bewildered many critics, and an explanation is needed. Lord Huang was related by blood to Wang Mang, the founder of the short-lived Xin Dynasty (9–23 AD). A self-proclaimed political reformer, Wang was commonly seen during

Liu's era as a wicked usurper of the sacred Han imperial household. Upon Wang's disgraceful death, Lord Huang took up a new surname and fled into seclusion in Yongzhou, where he practiced the Way of the ancient kings and won the respect of the natives. Regarding Liu's association of Lord Huang with Wang Mang, Xiaoshan Yang provides a convincing explanation. He believes that the story that ends the essay implies a moral lesson. We know that Liu was himself exiled to Yongzhou because of his involvement in the Wang Shuwen faction, which had by then been officially denounced as a group of usurpers. As Yang argues, "What Liu Zongyuan sees in Lord Huang is the possibility of moral redemption through the accomplishment of deeds rather than quibbling about words or empty names....Just as Lord Huang eventually succeeded in transcending the bad name of his ancestor, Liu envisions a morally optimistic prospect for reinstating his name in history through the tendering of practical service to the local people."[56]

To add to Yang's insight, I would further suggest that this moral lesson is encased neatly in a cartographic complex, thus embedding in the narrative a thinly veiled declaration of political salvation. The map-viewing strategy seen at the beginning of the essay may be considered an act of (re)connecting: the zooming-in effect refocuses visual attention on Yongzhou, lifting a place that appeared remote and irrelevant to everyday geographic experience to a spatial reckoning on a higher cartographic scale, and thus enlivening its link with the broader imperial world. The names of the ancient states and prefectures, strategically positioned in the four directions of the Chinese heartland, reconnects the unfamiliar place back to a history in which Lord Huang achieved redemption. This is the time-space in which the story of this local god is told. Here the author emerges as a free navigator, charting his course on the grand map as well as within political time, from his desolate place of exile back to the bustling imperial realm. Lord Huang's sagacious precedent serves as the source of confidence, a concrete way he might fulfill his lofty wish.

Although Liu never had the opportunity to reestablish his career in the imperial center, his reputation was reversed posthumously, just as Lord Huang's was. Like Lord Huang, Liu himself came to be worshiped as a local god after he died in Liuzhou in 819. His shrine was made into a cultural landmark, and it has enjoyed official endorsements from the Song Dynasty through the present. As with Lord Huang, Liu's legacy has become part and parcel of Yongzhou's historical heritage. The scenic sites he named came to define the cultural landscape of the prefecture for many centuries to come.

To conclude, Liu's poems and landscape essays discussed earlier show the clear literary navigation of a map-reading eye across the expansive imperial realm. These poems and essays show a comprehensive command of cartographic knowledge, which attests to his geographically conditioned epistemology. His literary mind traverses grand spaces, connects far-flung regions, and defies political misfortune with sharp historical consciousness. Throughout, Liu's poetic creativity lends further expressive power to the broad geographical exploration that was happening on several fronts during his era. Of course, cartography as a field of knowledge does not necessarily aim for the objective replication of geographical reality—nor would such a pursuit be a viable possibility. Rather, cartographical practices are themselves creative manipulations of the physical world. Liu's literary intervention went a step further to rearrange and synthesize the mediating cartographic perspectives, organizing them into poetic forms. In playing with both the exactitude and symbolism of distance, directions, and place names, he invented a new literary cartography through descriptive and metaphorical expressions. This adventurous literary feat in turn enabled him to negotiate space and time with the imperial center from a marginalized political position. As we have seen, this combination of literature and cartography worked in concert to push Liu Zongyuan, as a great stylist of both poetry and prose, to new heights of expressivity.

Map Viewing, Painting, and Artistic Creation in Zhang Hu's Poems

This section addresses the mid-Tang poet Zhang Hu's two painting-reading poems titled "Two Poems on Contemplating a Painting of the Mountains and Seas." A "Painting of the Mountains and Seas" is a classical landscape painting conventionally used to aid religious contemplation. In these poems, however, Zhang imposed a grand map view of the world onto the painting, and associates plants and animals in the painting with the peripheries of the empire. I argue that a cartographic perspective enabled the poet to dissociate the painting from its religious context and to infuse it with new imperial and artistic meanings. As an interface through which geography encounters visual art, Zhang's poems crystalized the transformative effects of the mid-Tang grand maps, themselves a creative combination of visual art and cartographic technology, on the viewers' experience with other forms of medial representation.

Alive between 792 and 849, Zhang Hu is conventionally considered a transitional poet from the mid-Tang to the Late Tang.[57] Although much of his work cannot be dated, we know that he had already established himself during the Yuanhe Reign (805–820). As some critics rightly point out, Zhang was an enthusiastic imitator of Li Bai, the high-Tang literary giant, in both his poetic style and his self-fashioning as a wild poet.[58] Zhang provides a perfect case for our discussion because he and his model Li Bai both wrote about a "Painting of the Mountains and Seas," but only Zhang wrote in an age of great advancements in cartographic knowledge. As I hope to show in the following pages, their different ways of engaging with the same genre of painting implies the powerful presence of map-viewing experiences in mid-Tang poetic recreations of the world image.

"Painting of Mountains and Seas" is a generalized name for paintings related to the ancient geographic work *Classic of Mountains and Seas.*[59] Based on textual records of now-lost paintings, we know that the customary images in the paintings include seas, mountains, immortal islands, and sea creatures such as whales and giant turtles. They present a fantastic world to aid Buddhist or Daoist meditation.[60] To begin our discussion, let us first read Li Bai's more famous "Viewing the Painting of Mountains and Seas in the Chamber of the Chan Master Ying" ("Ying chanshi fang guan 'Shanhai tu'" 瑩禪師房觀山海圖):

> This enlightened monk shut up his temple.
> Leaving no trace, he encompassed a broad mind.
> Cloudy mountains were painted on the screen.
> Amassed peaks entered the heavens.
> Looking at the densely tufted red scarps
> One wonders if they were folded curtains in daylight
> Penglai Island came to the window.
> The Ying Sea flowed into the tables.
> Misty surges contended with sprays.
> Islands and islets created a chaos.
> Journeying sails floated in the air.
> Waterfalls splashed across the sky.
> It seemed that the precipitous mountains could be scaled,
> But this fancy only left us sighing.
> Reaching the deep, the painting shared the darkness
>   with the true mind,
> So it suited the enjoyment of a man of quietude.
> Like climbing up to Red City,
> And high-stepping close to Cangzhou,
> Viewing it was a pleasure,
> And brought about much repose.

> 真僧閉精宇
> 滅跡含達觀
> 列嶂圖雲山
> 攢峰入霄漢
> 丹崖森在目
> 清晝疑卷幔
> 蓬壺來軒窗
> 瀛海入几案
> 煙濤爭噴薄
> 島嶼相凌亂

征帆飄空中
瀑水灑天半
崢嶸若可陟
想像徒盈嘆
杳與真心冥
遂諧靜者玩
如登赤城裡
揭涉滄洲畔
即事能娛人
從茲得消散[61]

This poem describes the speaker's internal experiences in a Chan master's chamber, mediated by an example of a "Painting of Mountains and Seas" that was hung in the room. In a religious space, the painting blurs the boundary between the imaginary world and the domestic reality, and it invites the speaker to embark on a journey to a transcendental realm populated by both natural images and mythical place-names. In the couplet "Penglai Island came to the window. / The Ying Sea flowed into the tables," these two mythical places belonging to the Daoist world are merged with the objects in the room, as if the entire chamber is transformed into a Daoist realm.[62] The poem then depicts the waves, islands, sails and waterfalls one after another, all of which are typical images of a "Painting of the Mountains and the Seas."

Following these landscape depictions, the poem combines a celebration of the painting's power with compliments to its owner for being one with "the true mind" and a "man of quietude." Toward the end of the poem, two more place names appear, Red City (Chicheng 赤城) and Cangzhou 滄州. Red City may refer specifically to the peaks of Tiantai Mountain, where Daoist culture flourished, or more generally to a mythical Daoist realm.[63] Cangzhou is a reference to a hermit's abode.[64] The line "Like climbing up to Red City, / And high-stepping close to Cangzhou" therefore suggests that the speaker is now completely removed from the physical environment of the chamber and is carried to these religious sites by his meditation.

The poem ends with a remark of satisfaction: the speaker's mental journey through the painted world is both pleasurable and relaxing, as is his visit to the Chan master's chamber. As a whole, the poem plays with spaces both lived and imagined. The place names it invokes belong to a mythical realm, unrelated in any way to localities in the real world. In other words, Li Bai's mental journey through this "Painting of Mountains and Seas" is almost purely religious, starting from the Chan master's chamber, journeying through mystical Daoist destinations, and then ending in a state of deep, satisfying meditation.

In contrast, Zhang Hu's "Two Poems on Contemplating the Painting of Mountains and Seas" derives significantly different imagery from a similar painting, even as he borrows from his predecessor's writing techniques. The first of the two poems goes as follows:

> In the ancient colors I examine what is blurry.
> The Chinese and the foreign are presented in the same hall.
> Glowing clouds open the caisson ceiling.
> Earth and heaven emerge from carved pillars.
> Slim stalks of grass grow out of tiny shoots.
> Light breezes brush dark hues.
> The mountains at night still hold their splendors.
> The autumn trees will not wither from the frost.
> Hidden bamboos have just shared some green.
> Dense blossoms are about to drip their fragrance.
> The sail lifts itself without any wind.
> Birds hover across the sun in the sky.
> Boats ride into the surf as if slowly engulfed by whales.
> The exhaled mirage of clams accumulates into the shape of towers.
> Now look at the mirror on the platform,
> Where all creation falls into its center.

> 古色辨微茫
> 華夷在一堂
> 雲霞開藻井
> 天地出雕梁
> 細草生毫末

輕風拂黛光
夜山猶帶景
秋樹不凋霜
隱竹纔分翠
穠華欲墮香
無風帆自起
度日鳥空翔
舟勢鯨吞久
樓形蜃吐長
更看臺上鏡
造化落中央[65]

The first two lines announce the organizing principle of the poem through a contrast between the mystical and archaic "ancient colors" (*guse* 古色) in the painting and a contemporary worldview. As the second line shows, the painting brings the Chinese and the foreign together in a "hall," an ordinary, nonspecific, and communal setting instead of a religious enclosure. The close co-existence of "the Chinese and the foreign" is both reflective of mid-Tang geopolitical realities and imitative of a world image as shown on grand maps. In this line, the poet thus intervenes by imposing a contemporary imperial vision on an ancient painting. In the following lines, in consequence, the imperial vision reframes the painting's conventional natural images by precluding their associations with the supernatural lands and allowing them to assume an artistic subjectivity.

An artistic appreciation of the painting then becomes the driving force of the poem. The poem is focused on a series of haunting illusions that the painting creates for the viewer, including the sail that "lifts itself without any wind," the birds that "hover across the sun in the sky," boats that "ride into the surf as if slowly engulfed by whales," and a mirage exhaled by clams that "accumulates into the shape of towers." Here, nature is detached from the seasons and becomes timeless. The sail and the birds flutter eternally; the whale and the clam are suspended in their movements. In the last two lines, the represented landscape appears so real that it merges with the well-ornamented house. Like John

Keats' Grecian urn with its frozen scene, the painting invites conjectures about the inner logic of an artistic world, arrests delicate moments in paradoxical language, and then concludes with the revelation of a certain universal truth. As Zhang envisions in the conclusion, it is creation, both in the artistic realm and in the geographic world of the empire, that lies at the heart of the painting. This, rather than religious reckoning, is the primal source of its images, lines, structures, and conceptions.

The second of the "Two Poems on Contemplating a Painting of Mountains and Seas" continues to entertain the same themes shown in the first poem with more geographic details:

> Whose brushwork ran wild
> And filled the whole wall with waves?
> The sun and moon lit the red arch.
> Mist and clouds rose from the painted roof-beams.
> Mountain fog opened the hues of morning.
> Sea vapor moved the aura of autumn.
> Occupying an island, a giant turtle opens its big eyes.
> Crawling through sand, a crocodile shows its long teeth.
> Pink waves light up Baiyue.
> Dark spots cluster in Sanxiang.
> Did I invoke meanings not in the painting,
> Getting lost in the realm of creation?
>
> 何人筆思狂
> 一壁盡滄浪
> 日月明丹拱
> 煙雲起畫梁
> 山嵐開曉色
> 海氣動秋光
> 島踞鰲睛大
> 沙行鱷齒長
> 粉波明百越
> 黛點簇三湘
> 詎作無圖意
> 空迷造化鄉[66]

The first half of the poem, up to "sea vapor moved the aura of autumn," elaborates on how the painting can transform a hall into an artistic world of rich sensuality. These lines are elegant variations of the same theme of artistic illusions that we saw in the first piece. The poem then takes a different direction, turning towards fauna and geography: "Occupying an island, a giant turtle opens its big eyes. / Crawling through sand, a crocodile shows its long teeth. / Pink waves light up Baiyue. / Dark spots cluster in Sanxiang." Images of unusual animals break from the familiar landscape of mountain and sea in the preceding passage, leading the poem toward the exotic realm of the empire's southern periphery. Highlighted by pink waves, the *Baiyue* 百越 region is traditionally associated with non-Han ethnicities in the far south of China, which largely overlapped with Lingnan, a land that extends well into today's Vietnam. The emphasized *Sanxiang* 三湘, clustered with dark spots, refers to the then underdeveloped regions in present-day Hunan.[67]

The well-traveled Zhang Hu was familiar with both of the regions mentioned in the poem. The introduction of the two place names maps the scene, with its giant turtle and crocodile as local animals, onto the empire's southern peripheries. The implied viewer in the poem does not stand on the artistic border between the human world and the immortal world, as she/he does with Li Bai's poem. Rather, the viewer sees an imperial realm where Chinese land stretches across mountains and seas into exotic spaces occupied by non-Han ethnicities and foreign peoples. This realm merges with the represented landscape in a colorful and fine-grained texture. At the end of the poem, the poet admits that he invokes meanings extrinsic to the painting and wonders if he is getting lost in the realm of recreation. Like the end of the first poem, here too the poet draws the reader into a dual creative process involving both the art and the world. This is a moment of transcendence achieved not by religious contemplation, but by a reflection on the relationship between art and human geography.

These two poems by Zhang read as surprisingly "modern." One reason is that they have an artistic autonomy that takes the place of spiritual pursuit as the principal embodiment of truth. This shift of focus is obvious when we compare Zhang's poems with Li Bai's earlier work. Zhang's pieces remind us of Western authors from Matthew Arnold to Wallace Stevens who proposed to use art or poetry to replace religion. Keats makes this kind of substitution fundamental to his Romanticism as well.[68] Yet what more fundamentally distinguishes the two poems is the unique interplay among language, painting, and Zhang's cartographic imagination that refreshes the reader's perspective and perception. In particular, the imposition of a contemporary map view on an ancient painting creates the decisive tension that alienates the poet from the religious realm commonly associated with the painting. The cartographic perspective then draws the poet into the imperial space, which is both familiar and enchanting: familiar, because it has markers such as "Chinese and the non-Chinese" and actual place names; enchanting, because the enigma of religious experience is now replaced by a celebration of the creation that produces both nature and the arts.

Intriguingly, in Zhang's two poems, a map view summons both the artistic and geographic creator into the poetic realm. In the following, I argue that the reason for this can be found in the transmedia nature of the grand map itself. As we know, maps were only vaguely differentiated from paintings in premodern China, and Jia's grand map is especially known for this this lack of differentiation. In fact, historians focusing on the technicality of early cartographic devices see Tang grand maps in a rather negative light. The historian Ding Chao, for example, argues that the use of painting styles in Jia's map was technically backward in comparison to the more efficient method of adopting standard symbols and legends. These techniques had already appeared in a Han Dynasty map found in Mawangdui, and would later become fully developed in the Ming Dynasty map "The Enlarged Terrestrial Atlas" ("Guang yutu" 廣輿圖). The "Map of Chinese and Foreign Lands" was therefore, in his view, a monumental regression in the history of Chinese cartography.[69]

If we suspend this uniquely modern scientific judgment for one moment and return to the actual representational effects the painterly maps achieved at the time of their creation, we gain entrance to a very different, long forgotten realm of medieval world image making. This historicized cultural perspective reveals new views of the maps themselves and their implications. A grand map done in a painterly style offers aesthetic insights and stimulates refreshing artistic creation in its readers. It expands viewers' geographical imagination and affords them the rare ability to visualize remote landscapes unreachable by the medieval foot traveler. Two poems written during the tenth century illustrate this point. The first is He Ning's 和凝 (898–955) "Yangchuan" 洋川:

> Yangchuan appears on the "Map of Chinese and Foreign Lands,"
> So I know it borders on blue mountains and green rivers.
> With few official duties I visit Buddhist temples.
> Living close to the river I should buy a fishing boat.
>
> 華夷圖上見洋川
> 知在青山綠水邊
> 官閒最好遊僧舍
> 江近應須買釣船[70]

When the speaker sees Yangchuan (in modern Yang County in Shaanxi Province) on the "Map of Chinese and Foreign Lands," he immediately visualizes it in the conventional context of a landscape painting marked by "blue mountains and green rivers." From there, he starts to envision a semi-reclusive life in Yangchuan, including visiting the temples and going fishing. We are not sure whether Jia actually painted the mountains and rivers in blue and green on his map, or whether "blue mountains and green rivers" involves the poet's derivative imagination of the surroundings of Yangchuan. But we do at least know that the map was crafted in such a way that a viewer immediately associated it with a landscape painting, and this association yielded much inspiration for the viewer's artistic production and his sense of his own life choices.

The second illuminating poem in this context is Wu Qiao's 伍喬 (fl. 940s) "Observing the 'Map of Chinese and Foreign Lands'" ("Guan 'Huayi tu'" 觀華夷圖):

> Other people's hands cannot achieve this sophistication.
> Notice that all the dense brushwork issues from the heart.
> At first I distinguish different counties from minute details,
> then gradually I see the four seas on the map.
> Roads with passes incline to channel the momentum of Chu,
> The mountains of Shu stretch high and green into Qin.
> The tip of the brush reveals everything in the imperial domain,
> Thus I will hang it forever in my courtyard.
>
> 別手應難及此精
> 須知攢簇自心靈
> 始於毫末分諸國
> 漸見圖中列四溟
> 關路欲伸通楚勢
> 蜀山俄聳入秦青
> 筆端盡現寰區事
> 堪把長懸在戶庭[71]

In the poem, the map viewer's eyes follow an invisible brush. He first notices that "all the dense brushwork issues from the heart," and then starts his visual journey from the different countries distinguished by the end of the brush and continues to the four seas that then take shape. The word *haomo* 毫末 in the third line means "minute details," but it can also literally mean the end (*mo* 末) of a brush (*hao* 毫). The brush portrays the geographic features with a sense of mobility and colorfulness, as "roads with passes" are seen to "incline to channel" as the "the mountains of Shu stretch high and green into Qin." In the last couplet, the poet finally mentions the "brush" explicitly. Clearly, the map represents the world as vividly as a painting. It enables the poet-viewer to reside in faraway, romantic places, if only for a brief, imaginary moment. The map likely did not pinpoint the position of the Shu Mountains on the earth as accurately as a modern military map would, yet it did connect the gentle slopes and steep cliffs to the broader geographic configuration of the world, in a way that other representational devices of the time could not. In both aforementioned poems, then, the painterly quality of the maps is presented as a positive, an essential aspect that accomplished much in its sheer artistic expressiveness and imaginative appeal to the geography-savvy literati mind.

We now arrive at a better understanding of how Zhang Hu was able to envision a grand map out of the "Painting of Mountains and Seas." His experience reading a grand map altered his approach to visual art, which he then beautifully poeticized in textual form. The cross-media examination also informs us about the unique features of Tang cartography. The grand map of the mid-Tang was itself a masterpiece of brushwork; the cartographic dimension only adds a sense of geographic precision and connection. As Cordell Yee says of traditional Chinese cartography in general terms, "What is now considered to be cartography participated in the same 'economy' of representation as poetry and painting. Because of this participation, one would expect cartographic artifacts to fall within a representational complex accommodating both rigid adherence and no adherence to material forms."[72] In the mid-Tang, what Yee terms the "representational complex" is manifested in Zhang Hu's poems, in which two representations of landscape, one geographic and one artistic, are overlaid on top of each other. The overlap may seem troubling to us as modern observers obsessed with scientific precision. Yet in its time, it achieved as much in nurturing new geographical mindsets as it may have sacrificed in technical details.

Conclusion

This chapter restores some of the most unique world images in mid-Tang literature to their intermedia context, in which grand maps featured prominently and formed conversations with various poetic and artistic expressions. The loss of the maps today does not diminish the significant roles they once played in the culture of their time, and should not prevent us from inquiring into the influence they once had on creative minds. Already during the Rebellion, Du Fu envisioned a grand unifying imperial space by reading a regional map. In the mid-Tang, the densely painted sublime image on the map that Du Fu celebrated took on its full power in Jia Dan's grand map, and the image continued to be a source of inspiration for poets. Indeed, the confluence of maps and poetry has existed in many ages and cultures, given the inherent connections and intersections among spatial cognition, visuality, language, and figures. Following Du Fu, Li He, Liu Zongyuan, and Zhang Hu made this confluence vital and all the more visible. Their works reveal that maps served as an aid to journeys both real and visionary, and they afforded cartographic features new expressive and evocative power. These poets embodied the high creativity of the mid-Tang literati who aspired to explore and even to transcend the troubled world in which they found themselves.

Notes

1. Du, *Dushi xiangzhu* 杜詩詳注, ed. Qiu Zhao'ao 仇兆鰲 (Beijing: Zhonghua shuju, 1979), 11.905.
2. For a comprehensive study of the literary representations of Sichuan's cliff-paths in Chinese literature and in Tang poetry in particular, see Fang Zheng's 方正 *Shudao yu Tangshi* 蜀道與唐詩, MA thesis, Xibei University, 2014.
3. Although the painting-map that Du Fu wrote about is no longer available, there are some extant traditional painting-maps that represent both the landscape and place names of Sichuan. The most famous one is the *Shuchuan shenggai tu* 蜀川勝概圖 (Painting-Map of Sichuan's Wonderful Scenes along the Yangtze River), attributed to the Song Dynasty painter Li Gonglin 李公麟 (1049–1106), and now in the Freer and Sackler Galleries. Another painting-map that has drawn critical attention in recent years is an anonymous untitled piece from the Qing dynasty, found in the Library of Congress. This map-painting is named by modern scholars the *Shanjing Shudao tu* 陜境蜀道圖 (Painting of the Cliff-Paths that Connect Shaanxi with Sichuan). See Bi Qiong 畢瓊 and Li Xiaocong 李孝聰, *Shanjing Shudao tu Yanjiu* 陜境蜀道圖研究, in *Ditu* 地圖, 204.4, 45–50 and Feng Suiping 馮歲平, *Meiguo guohui tushuguancang Shanjing Shudao tu zaitan* 美國國會圖書館藏《陜境蜀道圖》再探, in *Wenbo* 文博, 2010.2, 33–42.
4. The memorial suggests that Emperor Suzong send elements of the pro-government forces across the Yellow River to attack An Qingxu's 安慶緒 (723–759) rebel army. See "Memorial Written for Prefectural Officer Guo to Submit the Map for Exterminating Remaining Enemies" 為華州郭使君進滅殘寇形勢圖狀, in *Quan Tang wen*, 360.3653.
5. Du Fu's poems about his journey to Qinzhou have been well studied by a large number of scholars. For two recent studies of how the poems represented significant changes in Du's geographic perception of the empire, see Qi Hehui 祁和暉, "Dufu Qinzhou shi ji xie xiyu sichou zhilu shouduan zhancheng shanchuan renwen fengmao" 杜甫秦州詩記寫西域絲綢之路首段棧程山川人文風貌, *Du Fu yanjiu xuekan* 杜甫研究學刊, vol.4, 2014: 16–21 and Tian Feng 田峰, "Du Fu cong Qinzhou dao Ba Shu Jing Xiang de dili ganzhi yu wenhua tiyan" 杜甫從秦州到巴

蜀荊湘的地理感知與文化體驗, *Zhongguo yunwen xuekan* 中國韻文學刊, vol.30, no.1, Jan. 2016: 11–17

6. "The province's map gives it charge of Tonggu, / the post-station road leads to Drifting Sands. / Federate nomads join in a thousand tents, / the inhabitants make up ten thousand households. / The horses mettlesome, beads of sweat drip; / doing Turkish dances, whitened foreheads incline. / Young men, the lads of Lintao, even boast about coming from the west" 州圖領同谷，驛道出流沙。降虜兼千帳，居人有萬家。馬驕珠汗落，胡舞白題斜。年少臨洮子，西來亦自誇. For both the original text and translation, see Stephen Owen, trans., *The Poetry of Du Fu* (Boston: De Gruyter, 2016), book 7, 134–135.
7. See Zhang Zhongcai 張仲裁, *Tang Wudai wenren rushu kaolun* 唐五代文人入蜀考論 (Beijing: Shehui kexue chubanshe, 2013), 18–30.
8. As *Old History of the Tang Dynasty* comments, "the conception and style of his writing creates a feeling of high rocks and steep hills rising up to ten thousand *ren*. Many contemporary men of letters imitated it, but nothing could approximate it" 其文思體勢，如崇巖峭壁，萬仞崛起，當時文士從而效之，無能髣髴者. See *Jiu Tang shu*, 137.3772. *Ren*, also called *zhang*, is a Chinese unit of length. Ten thousand *ren* equals approximately 30,000 meters.
9. For some critical perspectives on this poem, see Wu Qiming 吳企明, 李長吉歌詩編年箋注 *Li Changji geshi biannian jianzhu* (Beijing: Zhonghua shuju, 2012), 6.723–724.
10. Ibid., 6.724.
11. Based on J. D. Frodsham's translation with modifications. See Frodsham, *The Poems of Li Ho* (Oxford: Clarendon Press, 1970), 31.
12. Li, *Li Changji geshi biannian jianzhu*, 6.721.
13. See *Ancient and Early Medieval Chinese Literature* (vols. 3 & 4): *A Reference Guide*, David R. Knechtges and Taiping Chang eds. (Leiden: Brill Academic Publishers, 2013), 1966–1967.
14. Tu Kuo-Ching, *Li Ho* (Boston: Twayne Publishers, 1979), 46.
15. Ibid., 1967.
16. *Jiu Tang shu*, 138.3785.
17. Pankenier, *Astrology and Cosmology in Early China* (Cambridge: Cambridge University Press, 2013), 299.
18. For example, Li Jifu says in the preface to *Maps and Treaties of the Provinces and Counties of the Yuanhe Reign*, "I have heard that the kings of the past established administrative regions and surveyed their territories. They observed order from the orbits of stars, and examined

the law in patterns on the earth." 臣聞王者建州域，物土疆，觀次于星躔，察法于地理. See Li, *Yuanhe junxian tuzhi* (Beijing: Zhonghua shuju, 1983), 1. Du You 杜佑 (735–812) also mentions the *Tongdian* 通典, "All the divisions of the country match the celestial phenomena above. This method originated at the end of the Zhou Dynasty, which established thirteen divisions to distinguish the regions on the earth" 凡國之分野，上配天象。始於周季，定其十二，其地可辨. See Du, *Tongdian* (Taipei: Xinxing shuju, 1963), 172.912. Another example is found in Liu Yuxi's "Records of the Hall of the Prefect of Lianzho" 連州刺史廳壁記. In it the author describes that "in terms of astrology, this prefect shares its celestial division with Jingzhou, and its border interlocks with that of Panyu" 此郡於天文與荊州同星分，田壤制與番禺相犬牙。 See *Liu Yuxi quanji biannian jiaozhu*, 15.1011.

19. Shrinking the earth (*suodi* 縮地), or shrinking the veins of the earth (*suodimai* 縮地脈), is a Daoist technique of miniaturization that the Daonist adept Fei Changfang is said to have mastered. See Robert Ford Campany, *To Live as Long as Heaven and Earth: Ge Hong's Traditions of Divine Transcendents* (Berkeley: University of California Press, 2002), 164.
20. Cao Yin曹寅 et al., *Quan Tang shi* 全唐詩 (Beijing: Zhonghua shuju, 1960), 716.8225.
21. De Weerdt, "Maps and Memory: Readings of Cartography in Twelfth- and Thirteenth-Century Song China," 156.
22. Needham, *Science and Civilization in China*, vol. 3, 545. *Quan Tang wen*, 828.6336.
23. Ibid., 972. It has been a consensus among scholars of Chinese cartography that Jia applied this method to the map. However, recent scholars Xin Deyong and Ding Chao maintain that previous scholars have misread Jia's memorial to Emperor Dezong. In fact, the two colors were used for the texts in *Notes on Administrative Jurisdictions and Foreigners in Four Directions*. Even were this the case, the added historical dimension to the map reading experience should not be denied, given that the texts were supposed to be available for the reader to consult when looking at the map.
24. *Jiu Tang shu*, 138.3786.
25. *Quan Tang wen*, 828.6336.
26. Lü, "Han yudi tu xu" 漢輿地圖序, *Quan Tang wen*, 628.6335.
27. See "Thirteen Poems from My Southern Garden" 南園十三首, *Li Changji geshi biannian jianzhu*, 4.512. The English translation is based on J. D.

Frodsham's translation with modifications. See Frodsham, *The Poems of Li Ho* (Oxford: Clarendon Press, 1970), 61. This poem is estimated to have been written in 813, while "On Leaving the City and Parting from Zhang Youxin I pledge to Li Han with Wine" was written in 812. See Li, *Li Changji geshi biannian jianzhu*, 4.461; 4.528.

28. Ibid., 4.454.
29. For more about *hetu*, see Anna Seidel, *Imperial Treasures and Taoist Sacraments: Taoist Roots in the Apocrypha*, in Michel Strickmann, ed., *Tantric and Taoist Studies in Honour of R. A. Stein*, vol. 2 (Bruxelles: Inst. Belge des Hautes Études Chinoises, 1983), 298–301.
30. Based on J. D. Frodsham's translation with modifications. See Frodsham, *The Poems of Li Ho*, 233–234.
31. Li, *Li Changji geshi biannian jianzhu*, 3.412.
32. Ibid., 3.415.
33. Ibid., 414.
34. *Record of Local Customs* (*Fengtu ji*) was a book of geographic writing composed by Zhou Chu 周處 (236–297) of the Jin Dynasty. See Liu, *Liu Zongyuan shi jianshi* 柳宗元詩箋釋, annotations and commentaries by Wang Guo'an 王國安 (Shanghai: Shanghai guji chubanshe, 1998), 3.367.
35. *Tong* 峒 literally means caves, and by extension it means the areas where ethnic minorities such as the Miao People lived. Ibid., 367.
36. The Zhou Documents 周書 refers to the *Remainder of the Zhou Documents* (*Yi Zhoushu*, 逸周書), a collection of seventy chapters compiled during the Western Han period. "The Meeting with the King" is the fifty-ninth chapter, and it describes a meeting of the King of Zhou in the capital with the envoys of sixty-one tribes with their various offerings of tribute.
37. Ibid., 3.366.
38. See Liu, *Liu Zongyuan shi jianshi*, 3.366–367.
39. According to the *Guliang Commentary* 穀梁傳, "The states in the center are called the Qi and the Song. The faraway states are called the Jiang and the Huang." 中國稱齊、宋，遠國稱江、黃. Ibid., 366–367.
40. Liu, *Liu Zongyuan ji jiaozhu*, 42.2881.
41. See Liu, *Liu Zongyuan shi jianshi*, 3.355.
42. Wang Qianjin 汪前進, "Xiancun zui wanzheng de yifen Tangdai dili quantu shuju ji" 現存最完整的一份唐代地理全圖數據集, *Ziran kexue shi yanjiu* 自然科學史研究, vol. 17, no. 3, 1998, 273–288.
43. The four cardinal directions included north 北, south 南, east 東, and west 西, often modified by *zheng* 正 ("due"); the four intermediate direc-

tions were: *dongnan* 東南 (southeast), *xinan* 西南 (southwest), *xibei* 西北 (northwest), and *dongbei* 東北 (northeast); the eight secondary-intermediate directions were: *zhengdong weibei* 正東微北 (east by northeast), *zhendong weinan* 正東微南 (east by southeast), *zhengnan weidong* 正南微東 (south by southeast), *zhengnan weixi* 正南微西 (south by southwest), *zhengxi weinan* 正西微南 (west by southwest), *zhengxi weibei* 正西微北 (west by northwest), *zhengbei weixi* 正北微西 (north by northwest), and *zhengbei weidong* 正北微東 (north by northeast).

44. *Yuanhe junxian tuzhi*, 37.926.
45. Mayang is roughly north of Liuzhou, slightly towards the west. Li Jifu describes Chang'an as straight to the north of Liuzhou, but to the northeast of Jinzhou, indicating that Jinzhou is to the northwest of Liuzhou. See *Yuanhe junxian tuzhi*, 30.749.
46. Based on Xiaoshan Yang's translation in "Naming and Meaning in the Landscape Essays of Yuan Jie and Liu Zongyuan," *Journal of the American Oriental Society*, vol. 120, no. 1 (Jan. – Mar., 2000), 94.
47. Liu, *Liu Zongyuan ji jiaozhu*, 29.1879 – 80.
48. See Zhang Shizhao 章士釗, *Liuwen zhiyao* 柳文指要 (Beijing: Zhonghua shuju, 1971), 838–839.
49. He, *Yimen dushuji* 義門讀書記 (Beijing: Zhonghua shuju, 1987), 646.
50. Yang, "Naming and Meaning in the Landscape Essays of Yuan Jie and Liu Zongyuan," 94–95.
51. Jin is roughly now Shanxi Province. Since the choronym (*junwang* 郡望) of Liu's family clan was Hedong (now Yuncheng in Shanxi Province), Liu often evoked Hedong and Jin in his writing and even discussed the cultural geography of the Jin area at length in his essay "Inquiry into Jin" 晉問. Yet Liu had never been to the Jin area in his life. Bin is today's Bin County 彬縣 in Shaanxi Province, and in the mid-Tang it was officially named Binzhou 邠州. When Liu visited this place in 794, it was on the northwest border of the Tang due to the radical shrinking of the empire's territory since the Rebellion. Wu generally refers to today's Jiangsu and Zhejiang provinces, which Liu had never visited. The eastern extreme of Liu's travel was Jiujiang 九江 (now Jiujiang in Jiangxi Province), which historically belonged to the Wu Kingdom, but it did not represent the typical landscape of the Wu area. The area between Chu and Yue is now the area in the intersection of Hunan Province, Guangdong Province, and Guangxi Province. For more about Liu's travel to Binzhou and Jiujiang respectively, see Shi Ziyu 施子瑜, *Liu Zongyuan nianpu* 柳宗元年譜 (Wuhan: Hubei renmin chubanshe, 1958), 15.11.

52. Sima Qian, *Records of the Grand Historian of China: Translated from The Shih Chi (Volume II: The Age of Emperor Wu 140 to Circa 100 B.C.)*, Trans., Burton Watson (Columbia University Press, 1971), 253.
53. See *Shiji* 史記 (Beijing: Zhonghua shuju, 1959), 116.2991.
54. See *Bai Juyi ji jianjiao* 白居易集箋校, Zhu Jincheng ed. (Shanghai: Guji chubanshe, 1988), 43.2764.
55. See *Tang wen shiyi* 唐文拾遺, ed. Lu Xinyuan 陸心源 (Taipei: Wenhai chubanshe, 1962), 22.340.
56. Yang, "Naming and Meaning in the Landscape Essays of Yuan Jie and Liu Zongyuan," 95.
57. For Zhang's life as a mid- to Late-Tang poet (中晚唐詩人), see Yin Zhanhua's preface to *Zhang Hu shiji jiaozhu* 張祜詩集校注 (Chengdu: Bashu shushe, 2007), 1–29.
58. Ibid., 4–7. See also Stephen Owen, *The Late Tang: Chinese Poetry of the Mid-Ninth Century* (827–860) (Cambridge: Harvard University Asia Center, 2006), 236–239.
59. Some scholars believe that the original "Painting of Mountains and Seas" preceded *Classics of Mountains and Seas* and the latter was in fact a textual description and annotation of the former. However, the original paintings had been all been lost before the Western Han. In the Han Dynasty, a new set of paintings were produced based on *Classics of Mountains and Seas* and the famous poet Tao Yuanming 陶淵明 (369–427) wrote a poem about them. The artist Zhang Sengyao 張僧繇 (fl. 6th century) produced a series of paintings titled "Pictures for Classics of Mountains and Seas" 山海經圖. It is possible that there were even more paintings titled "Paintings of Mountains and Seas" in circulation in the Tang dynasty. See Shen Haibo 沈海波, "Lun 'Shanhai tu' chansheng de niandai" 論山海圖產生的年代, *Shanghai daxue xuebao* 上海大學學報, vol. 9, no. 1 (Jan. 2002), 22–27; Sun Zhizhong 孫致中, "Shanhai jing yu Shanhai tu" 論山海經與山海圖 *Hebei xuekan* 河北學刊 (1987.1), 57–58.
60. For a recent discussion of the themes and images in the *Shanhai tu*, see Chen Yuquan 陳昱全, *Bei Song yuzhi mizangquan banhua yanjiu* 北宋御制密藏诠版畫研究 (M.A. thesis, National Taiwan Normal University, 2009), 104.
61. *Zengding zhushi Quan Tang shi* 增訂注釋全唐詩, Chen Yixin 陳貽焮 et.al. (Beijing: Wenhua yishu chubanshe, 2001), 172.1458.
62. In the Daoist world, the Penglai Island (Penghu 蓬壺) is where the immortals live, and the Ying Sea (Yinghai 瀛海) generally refers to the vast sea surrounding the human world. In Meng Haoran's "Having a

Banquet with Wang Changling in Daoist Wang's Study" 與王昌齡宴王道士房, we see a similar depiction, with Penghu referring to the same Penglai Island. Here is part of that poem: "The bookshelf curtain covered the immortals' sacred books. / The screen was painted with the *Shanhai tu* / Contemplating it after drinking wine like rosy clouds, / I felt as though I'd arrived in Penghu" 書幌神仙籙，畫屏山海圖。酌霞復對此，宛似入蓬壺. *Quan Tang shi*, 159.1622.

63. *Li Bai da cidian* 李白大辭典, Yu Xianhao 郁賢皓 et.al. (Nanning: Guangxi jiaoyu chubanshe, 1995), 262.
64. *Zengding zhushi Quan Tang shi*, 72.587.
65. Zhang, *Zhang Hu shiji jiaozhu*, 9.424.
66. Ibid., 425.
67. Zhang once served as magistrate of Nanhai County in Lingnan and his poems such as "Traveling in Xiang" 湘中行 indicate that he traveled deep into Hunan. See Yin Zhanhua's introduction to Zhang's life and works, in *Zhang Hu shiji jiaozhu*, 6.
68. Matthew Arnold proclaimed that "most of what now passes with us for religion and philosophy will be replaced by poetry." See Arnold, "The Study of Poetry," in William Harmon ed., *Classic Writings on Poetry* (New York: Columbia UP, 2003), 464. Wallace Stevens' poetry implies that poetry's "supreme fiction" should replace the supreme being of religion. See B. J. Leggett, "Stevens' Late Poetry," in John N. Serio ed., *The Cambridge Companion to Wallace Stevens* (Cambridge: Cambridge UP, 2007), 69.
69. See Ding, "Tangdai Jia Dan de dili (ditu) zhushu jiqi dituxue chengji zaipingjia" 唐代賈耽的地理（地圖）著述及其地圖學成績再評價, in *Zhongguo lishi dili luncong* 中國歷史地理論叢, vol. 27, no. 3, 2012.7: 154.
70. See He Ning, "Yangchuan" 洋川 (The Yang River), in *Quan Tang shi*, 735.8400.
71. See Wu Qiao 伍喬, "Guan Huayi tu" 觀華夷圖 (Viewing the *Huaiyi tu*), in *Quan Tang shi*, 744.8462.
72. See Yee, "Cartography in China," 137.

Chapter 3

The Shifting Shapes of the Local Sphere

Map-Guides and Literary Writing

Narrowing down the focus from chapter 2, this chapter will examine individual local spaces within the empire. As mid-Tang writers traversed the realm in the capacity of imperial officials, they wrote as much in far-flung regional outposts on the frontiers as in major metropolitan centers, giving the literature of the era a distinctively local flavor. Much has been written about this fascinating body of literature, especially those works that deal with the south.[1] Yet few scholars have explored the epistemological basis of local literary production in this unique period of geographic knowledge production. This gap in scholarship is not an oversight, but rather a reminder of how painfully little we know about the medieval world of more than thirteen centuries ago.

In this chapter, I bring in archaeological discoveries and recent textual scholarship on Tang map-guides in hopes of better understanding how mid-Tang writer-officials worked and wrote in their local posts. Specifically, I argue that local literary production of the mid-Tang likely involved a process I call cross-field intertextualizing. That is, a diverse body of literary texts grew in parallel with the blossoming of the production of local geographies. Literary texts sprouting from all corners of the realm tended to share certain similar narrative strategies, spatial imaginations, and cultural elements with local geographic works. While we cannot say for certain to what extent literary authors borrowed directly from local geographic works, the intershading between literature and geography in local places still hints at an emerging world of new spatial imaginaries. They collectively constituted an important layer of the cultural context in which mid-Tang local texts were produced. The purpose of this chapter, therefore, is to reconstruct this intriguing intershading between the two fields in the local space, thus offering new ways to read and interpret some of the less studied mid-Tang texts and genres in a richly cross-field context.

Among the local geographic works interacting with literary creation were the map-guides. As a genre, map-guides occupied an intermediate space in the Tang imperial system of knowledge about the geographic world. On the one hand, they were imperially commissioned packages of geographic information intended to aid regional administrators in local governance. On the other hand, this geographic genre differed from the comprehensive national geographical records and grand maps because it was centrally conceived but locally produced. Much like the local gazetteer (*difang zhi* 地方志) that came after it, it was usually compiled by the learned men of letters within a given prefect or county.[2] It often included information that may not have been considered meaningful by the imperial center, but was revered in the local sphere. As was mentioned in chapter 1, Li Jifu, the mid-Tang chief minister who compiled the celebrated *Maps and Treaties of the Provinces and Counties of the Yuanhe Reign*, famously complained that local legends and strange mysteries collected by local geographers were of no use to imperial geographic projects. For the new officials, however, the "useless" information that filled the pages of the map-guides—indigenous legends, strange stories, prose and poetic lines eulogizing local landmarks, and the like—were precisely what enabled them to enter local worlds of knowledge and meaning. Such complexity makes the map-guide, an age-old minor geographic genre that nonetheless proliferated in the Sui and Tang, a versatile intellectual resource that could be used to mediate between the imperial and the local, the literary and the political.

While map-guides had served as an incidental source of information for local-sphere writers since the very beginning of the Tang, it was after the An Lushan Rebellion that they became a notable presence in literature.[3] Well-known writers like Yan Zhenqing 顏真卿 (709–784), Yuan Jie 元結 (723–772), Bai Juyi, Liu Zongyuan, and Liu Yuxi referenced them explicitly in their writings. Few writings by these and other authors gave specific sources for their geographic information, but the level of accuracy in their local geographic descriptions suggests that their authors must have consulted local geographic works extensively. For reasons I will soon discuss, map-guides were likely a frequent and immediate recourse as the writers foraged for informational materials to support their creative tasks. This observation points to an interesting possibility: underneath the colorful surface of mid-Tang local literature was an epistemological layer of geographic knowledge and spatial imagination, the constitution of which was in part premised upon the wide utilization of map-guides.

This is certainly not to suggest that map-guides were a universal source of geographic information for all forms of literary engagement within any given local space. Rather, Tang writers referenced the map-guide most often in their writings in and about a cultural-political space I call the local sphere. I think of the local sphere as the aggregate of the social practices, political values, cultural knowledge, and spiritual traditions that define the difference between the actualities of the counties and prefectures and the central scheme of imperial uniformity. These could include the worship of local deities, the circulation of local legends and mysteries, the unique cultural conventions and social customs of local communities, and so on. The presence of the local sphere serves as a constant reminder of the scale of the empire and its vast interior; it embodies the diversity and unevenness that defined the cultural experience of the traveler-writers. While the grand imperial view as discussed in chapter 2 at once encompasses and differentiates the Chinese and their ethno-cultural others, local spheres constituted the building blocks of the Chinese empire. A centrally appointed official had to govern simultaneously for the empire and within the local sphere, performing locally appropriate rituals to establish himself as the shepherd of the locality. This bond of duty inspired mid-Tang writer-officials to produce a wide range of writings in and about the local sphere.

Although they were not the only source of geographic information available to mid-Tang authors in the local sphere, the map-guide had characteristics that most other works of the time did not. In the absence of new archaeological findings, we will probably never be able to say with absolute certainty how medieval writers collected and organized empirical local knowledge in the writing process. However, the distinctive merits of the map-guide suggest that they could have been a very useful and convenient first point of consultation for writer-officials eager to learn about the local place. Some Tang encyclopedias, or *leishu*, such as *A Primer for Beginners* (*Chuxue ji* 初學記) and *Classified Extracts from Literature* (*Yiwen leiju* 藝文類聚), also contain sections on regional geography, which could provide literary writers with both geographic knowledge and stylistic inspiration. Nevertheless, these encyclopedias were conceived of as comprehensive collections of imperial knowledge in all fields. Their ambitious coverage meant that the information they could include in each field and on each specific locality had to be sketchy and selective. In addition, encyclopedia editors prioritized immediate usefulness for literary composition in selecting traditional geographic texts, and thus would not include a large amount of updated geographical knowledge of a given place. The local map-guides, in contrast, can be seen as locally compiled encyclopedias of the specific cultural and political geography of the place concerned. The wealth of information they offered about a prefecture or a county far exceeded that found in general encyclopedias. As I will show in the close reading that follows, these map-guides may have informed or otherwise interacted with many mid-Tang local literary texts.

Reading local-sphere writings in light of the map-guide will, as I hope to show in the rest of the chapter, allow us to pierce through the literary layer of the local sphere and see the underlying geographic layer with which it was deeply enmeshed. This layer of subtext offered the literary texts both raw information and literary inspiration, sometimes resulting in striking parallels between the two layers in style and structure. In addition, with those authors who made their reliance on the map-guides explicit in their literary work, this act of mentioning in itself adds a sense of authenticity and authority, thus strengthening the political implications embedded in the literary texts. My new cross-field reading strategy also enables us to rescue some of the largely neglected local literary genres from undue obscurity. The first such genres to come to mind are office inscriptions (*tingbi ji* 廳壁記), mostly congratulatory prose carved in the wall of a local official hall, and orisons of appeal (*zhuwen* 祝文), which are prose pieces dedicated to local deities for pragmatic purposes like rain making.[4] Both these genres of writing feature in the literary collections of important Mid-Tang writers. Yet they have been less studied as literary texts by modern scholars largely due to their pragmatic official purposes. My cross-field reading, however, shows that many such pieces were carefully crafted with rich embedded cultural-political meanings. Local writings were, after all, political in nature. They were composed by local offcials to strengthen local governance, conduct the grand imperial restoration, or convey personal or collective appeals to the court. These political implications were often masked by the rigid form, but were also reinforced by the authors' skillful appropriation of local geographic works.

In the literary works to be discussed in the following sections, map-guides are referred to as *tujing* or *tuji/tuzhi* 圖記/圖志 (map-records).[5] Occasionally, the author only mentions a map (*tu*), and in cases in which the context tells us that the map was accompanied by lengthy, detailed textual explanations and was furnished to the author to aid with his official duty, I treat the term as an abbreviation for map-guides.[6] As we are aware, for literary authors, ancient or modern, citing references was largely optional. Therefore, in addition to the texts analyzed in this chapter, there may have been many more works that drew upon map-guides without acknowledging the sources of information. This should come as no surprise, as geographical knowledge was not in itself the purpose of most of local sphere writings. It was an underlying subtext, and in most cases stayed in the background. What I have to offer may be just the tip of an iceberg, but given the scarcity of available sources so far, any estimate of the precise magnitude of this local literary phenomenon can come only as an educated guess.

In the following, I will first discuss Yan Zhenqing's two inscriptions for Daoist sites that explicitly mention the local map-guides. Written slightly earlier than the mid-Tang, Yan's inscriptions give us an overall picture of how the map-guide would inspire and facilitate literary writing in the post-Rebellion age. The map-guide points out local sites that embody local spiritual traditions, and leads Yan, the newly posted local official, to inspect some of them. Yan's mentioning and incorporation of the map-guide in turn helped him establish a sense of historical continuity and imperial authority in his writing. All of these dimensions reoccur in works by the mid-Tang authors. Following this, I will then move on to discuss mid-Tang writings in three parts, focusing on writings that appeared in government offices, engaged with local shrines, and recorded the sightseeing of a given place by visitors.

THE MAP-GUIDE AND YAN ZHENQING'S DAOIST INSCRIPTIONS

Widely known as one of the central figures of Chinese calligraphy, Yan Zhenqing was also a prominent political figure celebrated for his loyalty to the Tang court during and after the Rebellion.[7] A member of the noble Yan clan, which produced many cultural luminaries in medieval times, Yan Zhenqing earned his political reputation by staging a firm yet skillful campaign of resistance against the An Lushan rebels in his official post in the Shandong peninsula. With his steadfast dedication to the restoration of the Tang, he emerged as one of the most respectable court ministers in the post-Rebellion years. He was demoted to a series of regional posts in the south between 766 and 777 due to his misfortunes in factional politics, but continued to exert considerable political influence until his execution by the rebellious military governor Li Xilie 李希烈 (?–786) in 785. Yan's posthumous high reputation as a Confucian moral paragon was matched by his great influence as one of the most accomplished calligraphers in the Tang.[8] However, this fame also worked against him, overshadowing the literary value and intellectual complexity of his writings. It is only in recent decades that scholars of Daoism, cultural history, and the history of science have begun examining Yan's artistic creations, particularly several of his Daoist inscriptions, as serious intellectual texts.[9]

My discussion approaches the literary merits and politics of Yan's inscriptions from the perspective of their interaction with map-guides. The central focus of my analysis is Yan's two pieces of official writing that were inscribed in stone in Fuzhou (in present-day Jiangxi Province), where he served as prefect between 768 and 771. They are "Record of the Altar of the Immortal of Mount Magu, Nancheng County, Fuzhou" ("You Tang Fuzhou Nancheng xian Magu shan xiantan ji" 有唐撫州南城縣麻姑山仙壇記), and "Stele Inscription for the Altar of the Transcendent Lady Wei of the Jin Dynasty, the Lady of the Southern Marchmount, Primal Worthy of the Purple Void, Concurrently Supreme True Mistress of Destiny" ("Jin Zixu yuanjun ling Shangzhen siming Nanyue furen Wei furen xiantan beiming" 晉紫虛元君領上真司命南嶽夫人魏夫人仙壇碑銘). The first piece was written and inscribed for the ancient altar of Maid Ma before Yan left his post in Fuzhou in 771; the second piece, devoted to another Daoist immortal's altar in the same prefecture, was composed and commissioned for inscription a bit earlier, when Yan had just arrived at his Fuzhou post. Both pieces invoked the map-guides as a prominent source of information and inspiration, thus attesting to the three-way exchange among literature, geography, and the spiritual tradition in a politicized local space. As the following close reading will show, the map-guide at the same time affords the literary text epistemological and stylistic authority and anchors the historical continuity that the author was eagerly pursuing after a disruptive period of political chaos and human displacement.

That Yan Zhenqing, not an adherent of Daoism, wrote for Daoist altars while in Fuzhou would have come as no surprise to his Tang contemporaries.[10] The early Tang emperors worshiped Laozi, the ancient sage in the Daoism pantheon, as an ancestor in their royal bloodline. This early imperial strategy for legitimacy led the state to sponsor the religion and to elevate Fuzhou, an important site of Daoist legends, as a sacred locality in the realm before the Rebellion. When Yan eulogized local deities, he did so in the voice of an imperial agent who was helping to revive local Daoist culture for a larger imperial cause. This political background should be helpful as we work our way through the texts in the following pages.

Let us first look at the inscription for the altar of Mount Magu, a fascinating example of how the geographic layer of local knowledge enabled Yan to anchor the imperial cause in a vibrant local spiritual tradition through literary narration. In the beginning of the record, Yan quotes the hagiography of Maid Ma from Ge Hong's (葛洪, 284–364) *Biographies of Divinities and Immortals* (*Shenxian zhuan* 神仙傳). The hagiography tells how Maid Ma manifests herself as a beautiful young woman to another immortal, Wang Yuan (courtesy name Fangping), at the home of Wang's human disciple. It is a story of other-worldly fantasy, aimed at creating a sense of wonder in the reader. Chatting with Wang, Maid Ma says, "since I entered your service, I have seen the Eastern Sea turn into mulberry fields three times. Crossing over to Penglai, the water came only up to one's waist. I wonder whether it will turn to dry land once again." Fangping then answers, "The sages all say that the Eastern Sea will once again raise dust"[11] ("*Jieshi yilai jian Donghai sanwei sangtian, xiangjian Penglai shui, nai qianyu wangzhe, huishi lueban ye, qijiang fu huanwei luling hu?" Fangping xiaoyue, "Shengren jieyan haizhong xingfu yangchen ye*" "接侍以來見東海三為桑田，向間蓬萊水，乃淺於往者，會時略半也，豈將複還為陸陵乎。"方平笑曰，"圣人皆言海中行复扬尘也.")[12] The conversation bears little resemblance to the real world. "Penglai Island in the East Sea" is a typical allusion to the mystical realm beyond human reach in mystery narratives of the time, and immortal Wang's light-hearted reference to sagacious teachings is nowhere to be found in the Chinese classics. In fact, the story was thought of as so much beyond human temporality that its motif was abbreviated into the popular phrase "seas and mulberry fields" (*canghai sangtian* 滄海桑田), a metaphor for the wild vicissitudes of human history.

In the next section of the record, however, Yan brings this fantasy to the human world—specifically to the local space over which he presided—through an imaginative deployment of local geographical icons. The text here is both vivid and precise, displaying considerable literary quality:

> In the third year of the Dali era [768], I was made governor of Fuzhou. According to the map-guide, the mountain of Maid Ma is in the Nancheng district. At the summit is an ancient altar. Tradition says this is where Maid Ma attained to the Way. To the southeast of the altar is a pool. In the center was a red lotus, which suddenly turned blue recently. Now it has become white. Below the north end of the pool, beside the altar, are firs and pines, all bent into a canopy. Occasionally the bell and chime sounds of [Daoist immortals] pacing the void have been heard. To the southeast is a waterfall, which cascades more than three hundred feet. To the northeast there is a monastery on a stone outcropping. There are still mussel and conch shells embedded in the high rocks, which some believe to be remnants of the transformations of the mulberry fields [into the Eastern Sea and back again over aeons]. To the northwest is the Ma River. This is perhaps the site in Xie Lingyun's [385– 433] poem, "I Go into the Third Valley of the Ma River Where Huazi Hill Stands." There is a divinity residing at the mouth of the spring, by whom prayers for rain are speedily answered.[13]

> 大曆三年，真卿刺撫州。按圖經，南城縣有麻姑山，頂有古壇，相傳云麻姑於此得道。壇東南有池，中有紅蓮，近忽變碧，今又白矣。池北下壇傍有杉松，松皆偃蓋，時聞步虛鍾磬之音。東南有瀑布，淙下三百餘尺。東北有石崇觀，高石中猶有螺蚌殼，或以為桑田所變。西北有麻源，謝靈運詩題入華子岡是麻源第三谷，恐其處也。源口有神，祈雨輒應。[14]

The passage transforms the Daoist legend of Maid Ma into an empirical fact observed by a surveyor's eye with a sense of wonder. In Ge Hong's hagiography, Maid Ma recalls the metamorphosis of the Eastern Sea into mulberry fields. In the record, Yan actualizes the Daoist fantasy when he points to the remains of mussels and conch shells visible in the rocks near the altar, suggesting that the very location of the altar might have been the seabed which turned into mulberry fields. He encapsulates this inferred parallel in a series of well-chosen images: the descending and ascending perspectives, the color-shifting lotus flowers, and the Daoist music that drifts in and out. All these local features that indicate the ever-changing nature of the environment correspond nicely with the fantastic "sea and mulberry field" metaphor. At the end of the passage, Yan goes on to introduce other geographical objects surrounding the altar, including the Ma River and the shrine of an anonymous rain god, thus firmly situating the altar within the physical geographical setting of the local space.

Yan's writing about the local space is intricately related to the map-guide. Indeed, the passage starts with the phrase "according to the map-guide" (*an tujing* 按圖經). It is clear that he used the map-guide as a geographical reference to locate and inspect the site, but since the map-guide of Fuzhou that Yan relied upon is lost, it is unclear whether the following description of the site was written by Yan upon his own examination of the area or whether he was quoting from the map-guide. Nevertheless, Hua Linfu, a historian of the map-guides, considered the entire passage to be primarily a quote or a paraphrase from the map-guide of Fuzhou.[15] In the following, I will compare Yan's passage with three extant fragments from other map-guides to further demonstrate the intertextuality between the two. The ways Yan describes the altar, engages with Xie Lingyun's 謝靈運 (385–433) poem, and introduces the rain god all bear striking similarities to the map-guides. The serial presence of all these structural affinities between the text and map-guides in the region suggest that the resemblance was no coincidence, but more likely the result of an intertextual cross-pollination.

The first point of comparison is from the map-guide of She County 歙縣. A county some 180 miles to the northeast of Fuzhou, She County also featured a Daoist altar on the famed Beiyi Mountain 北黟山 (now known as Yellow Mountain 黃山):

> On the main peak is the altar of the immortal Master Fuqiu, where colorful clouds and mythical birds perch. This was where Master Fuqiu and Master Rongcheng wandered. Once someone went to the altar and suddenly saw splendid buildings and platforms. In front of one of the buildings there was a lotus pond; around it there were piles of salt and rice. He went back to bring villagers to collect the food, but could not find the spot again. People living at the foot of the mountain often heard heavenly music coming from the peaks.
>
> 山中峰有浮丘公仙壇，彩霞、靈禽棲止其上，是浮丘公與容成子遊之處所。昔有人到壇所，忽見樓臺煥然，樓前有蓮池，左右有鹽積，米積，遂歸，村人上取，了不知其處所。山下人往往聞峰上有仙樂之聲。[16]

In this passage, the way the altar and its surroundings are depicted is strongly reminiscent of Yan's representation of Maid Ma's altar in Fuzhou. Legends and wonders of the local immortals augment the altar's mysterious appeal, the lotus pond adds to the site's enigmatic beauty, and the ceremonial music occasionally heard by those living in the human world below elevates the mountain to a higher realm of Daoist religiosity. The similarities between Yan's work and map-guides indicate that there was a fluid relationship between local writing and geographical representation of local religious sites in the region. Or, to restate it slightly, the similarity between the two texts suggests a porous boundary of the two fields of representation concerning the local sphere.

The second fragment of a geographical text is taken from the *Map-Guide of Yuzhang* (*Yuzhang tujing* 豫章圖經). Yuzhang is now Nanchang in Jiangxi, also close to Fuzhou. Just as Yan evokes the Six Dynasties poet Xie Lingyun's poem when introducing the Ma River, this map-guide entry quotes a different poem of Xie when depicting the Pine-Gate Mountain in its territory:

> Pine-Gate Mountain takes its name from its many pine trees. To the north of it are the Yangtze River and Pengli Lake. There is a stone mirror on the mountain, which radiates brilliant light. Xie Lingyun's poem "Entering Pengli Lake" describes the place thus: "Climbing the high-banked shore I looked at the stone mirror / Carrying leaves I entered the Pine-Gate."
>
> 松門山者，以其多松，遂以為名。北臨大江及彭蠡湖，山上有石鏡，光明照人。謝靈運《入彭蠡湖口》詩云，"攀崖照石鏡，牽葉入松門"，是也。[17]

In fact, quoting famous literary writing about a local site is common in map-guide compilation.[18] As discussed in the Introduction, both the compilers and the readers of the local map-guides in the Tang Dynasty were primarily literati. Matching the geographical space with well-known literary allusions helped the literati navigate through the local space with a familiarity with its history and culture. In particular, after the Rebellion, educated northern elites served in the south extensively and were especially attentive to the literary allusions from the Six Dynasties—a time of southern literary flourishing—recorded in the local map-guides.[19]

Lastly, the map-guides also typically involve a category for local shrines, under which one finds records such as the following:

> Master of Rain:

> The above is two *li* east of the prefectural city. A shrine was established to house an icon of the god. People pray to this god when there is drought. The founding date is unknown.
>
> 雨師神
> 右，在州東二里。立舍，畫神主。境內亢旱，因以祈焉。不知起在何代。[20]

Such shrines were a hallmark of local spiritual life. Yan's line "a divinity residing at the mouth of the spring, by whom prayers for rain are speedily answered" describes just such a local shrine for the rain god. Yan's shrine was certainly different from that contained in the map-guide entry. However, the empathetic introduction of the location and function of a local shrine in a literary text reflects a potentially common spatial conception of the local spiritual world that cut across the map-guide/literary text boundary.

The previous analysis demonstrates an affinity between Yan's work and the map-guides. Yan may have used one or more map-guides as a reference, or he may simply have composed his literary work in a style that was shared by authors of the local geographical work. In any case, this affinity allows us to read Yan's passage as born out of a broad context of regionally circulating spatial conceptions and imaginaries, which the map-guides helped construct and popularize. This then leads us to consider the influence this layer of geographical subtext may have had on the writing and reading of Yan's literary text. Depending on how Yan utilized them, the map-guides, when explicitly invoked, may have functioned as an etymological support to his literary writing, or served as a reminder to contemporary readers of the literary text's local empirical authority. Either way, the intershading between the two types of texts, explicitly expressed or implied, gave the literary text a sense of geographic authenticity, thus helping to tie the otherwise free-floating legends to specific geographic locations in the author's local place.

This form of literary localization then allows Yan to link Maid Ma's altar —now affirmed as a site of Daoist wonders—to the Tang Empire's recent history of promoting Daoism. Immediately following the description of the altar and the surrounding environment, Yan brings the image of Emperor Xuanzong into the picture:

> During the Kaiyuan era [713–742], the Daoist Deng Ziyang practiced the Way at this spot. He received an imperial summons to enter the Great Unity Palace, where he exercised his merit and virtue [as a Daoist *fangshi*, or doctor] for twenty-seven years. One day there suddenly appeared in the courtyard a dragon chariot drawn by tigers and two men holding tallies. Deng Ziyang turned and said to his friend Zhu Wuyou, "They have come to receive me. You may report to the throne for me that I wish my corpse to be returned for burial on my mountain." Then he asked that a temple be set up at the side of the altar. Emperor Xuanzong complied with this. In the fifth year of the Tianbao era [746], when the "tossing the dragon" ritual was performed in the waterfall, a yellow dragon was seen in the stone pool. Emperor Xuanzong was excited, and he ordered that repairs be made to the abode of the Transcendent [Deng Ziyang] and the icons of him, his attendants, and cranes with clouds.[21]
>
> 開元中，道士鄧紫陽於此習道，蒙召入大同殿修功德。二十七年，忽見虎駕龍車，二人執節於庭中。顧謂其友竹務猷曰：此迎我也。可為吾奏，願欲歸葬本山。仍請立廟於壇側，元宗從之。天寶五載，投龍於瀑布，石池中有黃龍見。元宗感焉，乃命增修仙宇、真儀、侍從、雲鶴之類。[22]

The story tells how a Fuzhou-based Daoist priest entered the emperor's service and manifested his mystic power in a supernatural encounter with the gods. The emperor followed the call of the heavens by granting the priest's final wish, and then showed his deference to divine powers again after witnessing wonders in an imperially sanctioned Daoist ritual (known then as "tossing the dragon" [*toulong* 投龍]).[23] By recreating the encounter between the Emperor and a local Daoist priest, the text transforms the local place of Fuzhou into a critical site where the two fundamental traditions of the empire—the spiritual and the political—encountered each other and blended into a grand harmony.

The next and final part of the text links the two worlds above with contemporary Fuzhou, where the local Daoist culture is represented as a long and continuous tradition, from Maid Ma's age to the present time:

> Alas! In the past, Maid Ma left her traces on this ridge [Mount Magu]. The "True Immortal of the Southern Peak" left her altar at Guiyuan. Miss Flower manifested her unusual qualities at Jingshan. In the present age, the female Daoist Li Qiongxian, is eighty years old, but her appearance is very youthful. After Zeng Miaoxing had a dream in which Li Qiongxian instructed her, she ate only flowers and gave up grain. Deng Ziyang's nephew is named Decheng. He continues the practices involving incense and paper money. His disciple Tan Xianyan reveres the writings on the Daoist arts, and Shi Yuandong, Zuo Tongxuan, and Zou Yuhua are all pure and empty in their service of the Way. If this locale is no different from the luminous numina of other regions, then why has there been such a splendid succession of worthies here? As I am fortunate to help carry on this remaining tradition, I have dared to engrave this record in stone. The time is the summer, the fourth month of the sixth year [of the Dali era].[24]
>
> 於戲！自麻姑發迹於茲嶺，南真遺壇於龜源，華姑表異於井山。今女道士黎瓊仙，年八十而容色益少，曾妙行夢瓊仙而飡花絕粒。紫陽姪男曰德誠，繼修香火；弟子譚仙巖，法籙尊嚴。而史元洞、左通元、鄒鬱華，皆清虛服道。非天地氣殊異，江山炳靈，則曷由纂懿流光，若斯之盛者矣？真卿幸承餘烈，敢刻金石而志之。時則六年夏四月也。[25]

By juxtaposing the Daoist immortals with living priests who carried on their spiritual legacy, the text magnifies the sense of historical and spiritual continuity that Fuzhou symbolized through its unbroken Daoist lineage. At the end of the record, the author presents himself as the guardian of this local tradition, a tradition that has by now gained greater political significance beyond the local sphere.

While there is no definitive evidence to show that Yan's text is linked directly to a particular map-guide, it displays a significant structural and spatial resemblance to the genre. The legend-based visualization of a Daoist religious space, the shared reference to a Six-Dynasties cultural icon, and the strategic positioning of local rain shrines at the notable margins of the representational landscape—the active presence of all these elements in the textual space demonstrates that Yan worked with spatial imaginaries and narrative strategies that also featured in the map-guides. A map-guide led the author to the physical site of his literary subject, possibly provided him with geographical information, descriptions, and allusions that he would need for his literary writing, and gave the writing a sense of authority. Thus, the map-guide helped Yan locate a spiritual tradition that embraced the ever-changing nature of the world in a localized geographical-literary space, and then to reframe it in a continuous cultural genealogy endorsed by imperial power. This political-literary function of the map-guides—as a collective body of work and as individual sources of information—emerges even more clearly in another of Yan's inscriptions, "Stele Inscription for the Altar of the Transcendent Lady Wei of the Jin Dynasty, the Lady of the Southern Marchmount, Primal Worthy of the Purple Void, Concurrently Supreme True Mistress of Destiny," to which we now turn.

The "Transcendent Lady Wei" in the title of the inscription refers to Wei Huacun 魏華存 (252–334), an important deity who was widely celebrated by Daoists from the early medieval age through the Tang. Just like the record of Maid Ma's altar, this inscription begins with a hagiography that affirms Lady Wei's presence in Fuzhou before her transcendence. Lady Wei had labored to set up sites of Daoist worship in the prefecture, but by the Tang dynasty, the physical traces of the sites were gone. The only evidence of their past prosperity were the textual records, one of which was extant in a map-guide:

> When the Lady [referring to Wei Huacun] first crossed the river, she travelled around the famous mountains and arrived in Linchuan Commandery [in Fuzhou in the Tang]. She erected an altar and established a monastery to the west of the Ru River. She then built a grave some one hundred steps to the east of the yard and set up an altar field at Mount Shijing. She traveled between

> them for pleasure and leisure. This was long time ago and [the constructions] have disappeared in the wild woods and grasses. Although the map-guide records all of them, there is not a single remaining trace.
>
> 初夫人既渡江，徧遊名山，至臨川郡，臨汝水西立壇置精舍，院東百餘步造冢壙，又於石井山建立壇場，往來遊憩。歲月深久，榛蕪淪翳，雖備載《圖經》，而略無遺跡。[26]

Yan then recounts how in the early Tang, the local female Daoist Huang Lingwei 黃令微 (640–721) managed to identify and rebuild all the sites, and how Emperor Xuanzong's patronage allowed them to prosper through the High Tang. However, when Yan inspected the sites as prefect in 768, he found a desolate scene:

> In the year of 768, I, Zhenqing, was appointed the governor of this prefecture. I revered the place in words and visited with proper etiquette. The misty, far-off suburban area and the outer city wall had been disturbed by thieves. The female priests were powerless and their servants had fled, while jackals and wolves gathered in nearby caves. It was as if the appearance of the immortal was still there, for the altar and the palace were still standing. I looked at them with respect and walked around, and was quiet as if something had been lost to me.
>
> 大曆三年，真卿叨刺是州，言崇禮謁。郊郭蒙邃，萑蒲震驚，女弱曹逃，豺狼窟聚。真儀如在，壇殿巋然，瞻仰徘徊，悄焉若失。[27]

Although Yan does not make it explicit, it was most likely the Rebellion and the ensuing political dysfunction that caused this decline. This unsettling picture of desolation and decline appears as an apt metaphor for an empire that had failed and faltered over a decade of disorder. Luckily, Yan was still able to identify the altar and the palace. As we have seen in Maid Ma's case, the map-guide led Yan to locate and inspect local religious sites. Here, again, Yan might also have sought help from the map-guide in order to "revere the place" and "visit with proper etiquette." In addition, the map-guide and the local geographic knowledge it offered also served as a metaphorical anchor for historical continuity and a resource for restoration. As the later parts of the inscription would reveal, Yan then collaborated with local Daoist leaders to renovate Lady Wei's altar and wrote an inscription to commemorate the site. As a result, Yan delineates in this inscription a temporal contrast specific to his own time, crafting a narrative of historical continuity that both resists and attempts to ameliorate the profound disruption that recent political changes had brought to the local place.

To conclude, if we take these two inscriptions together, the implicit interplay among the map guides, the inscription, and the religious site is revealed in a clearer light: the map-guide, as the textual form of local geography, spatializes and localizes the religious tradition through its descriptive narration. The inscription, as a materialized form of literature, draws upon spatial imaginaries also present in the map-guides to represent and reconstruct this local cultural geography. Such reconstruction was as much for greater imperial political purposes as for deep personal introspection. Yan's geographically inspired writing accompanied efforts to restore imperial glory after a turbulent era. At the same time, the localized geographic way of thinking also enabled Yan to see through the immediate history and reach a broader vision at the level of the constant human and natural changes in the longue durée, or of the cyclic transformations between seas and mulberry fields.

Map-guides and Local Office Spaces

In this section, I will discuss how map-guides helped mid-Tang writers articulate the local sphere through writings that would appear in local government offices. In these cases, map-guides continued to intertextualize with the local writings in question, sometimes functioning as a useful source of local knowledge and an anchor of imperial authority. Examples from two writers best illustrate this literary use of map-guides:

Liu Zongyuan's "Office Inscription for the Vice Magistrate of Wugong County" ("Wugong xiancheng tingbi ji" 武功縣丞廳壁記); and Liu Yuxi's poems "Seventy Rhymes to Express My Thoughts in Liyang" ("Liyang shushi qishi yun" 歷陽書事七十韻) and "Office Inscription of the Prefect of Hezhou" ("Hezhou cishi tingbi ji" 和州刺史廳壁記). Liu Zongyuan's text was written for his friend for an inscription in a county office, while Liu Yuxi's was written for his own office in Hezhou prefecture. Both authors either referred to or implied map-guides as a first point of reference for local knowledge. They restructured the diverse information from this and possibly other local sources into intricate literary narratives about the geography, culture and history of the local sphere, and imbued them with subtle political meanings. In the process, the geographical voice in the local sources was transformed into a literary articulation addressing a much larger political audience.

The major literary genre this section engages with is office inscription. This refers to stylized prose records inscribed on a wall to commemorate various governmental halls, composed either by the official serving there or by a friend or colleague who had literary talent and fame. Office inscriptions were in vogue during the mid-Tang and caught the interest of many prominent writers.[28] These inscribed texts often eulogized the defining geographic features and cultural merits of the region under their jurisdiction, creating a peculiar kind of geographic microcosm in the confined space of the office. Notably, the intertextual relationship between these official prose pieces and the map-guides went both ways. On the one hand, as the historian Xin Deyong has argued, local map-guides were among the first documents that an official had to carefully read upon arriving at his new post.[29] In cases where such officials composed their own office inscriptions, therefore, they would have had the information from the map-guides in the back of their minds as they wrote. On the other hand, as the mid-Tang writer Ma Zong 馬總 (fl. 785–823) succinctly explained in his "Office Inscription of the Prefect of Yunzhou" ("Yunzhou cishi tingbi ji" 鄆州刺史廳壁記), the office inscriptions were at the time viewed as a genre with geographical significance whose informational functions overlapped with those of the preceding map-guides:

> For the local customs and products, and the government's past accomplishments, one needs neither to inquire of the elderly nor consult documents and maps. When one enters the hall [and reads the inscription], everything can be discerned and understood.
>
> 其土風物宜，前政往績，不俟咨耆訪耋，搜籍索圖，一升斯堂，皆可辨喻。[30]

This passage demonstrates that the two genres of writing, office inscription and map-guides, represented geographic spaces of comparable scope, serving similar informational and intermediary purposes. The composition of either could take place in a context where the author was informed, inspired, or unconsciously influenced by the cultural elements and representational strategies of space in the other. In the three office inscriptions I will discuss next, the authors all indicated explicitly that they consulted map-guides before committing their thoughts to the wall. While this certainly does not exclude the possibility that they also took raw material from other sources—indeed, it is likely that they did—it lends substance to the hypothesis that these specific inscriptions intertextualized in significant ways with the map-guides. In the following, I will read these inscriptions in light of this hypothesis to analyze how a local geographic subtext may have been at work in the literary-political utterances of the local writer-officials.

First let us look at Liu Zongyuan's inscription, which he composed for a vice magistrate of Wugong County in 802.[31] In it, Liu twice mentions his consultation of a county map. The map first appears when Liu introduces a highly condensed historical and cultural geography of Wugong County:

> Wugong is a large county in the greater capital area. Having consulted its map, [I find that] the land, then called *youtai,* was bestowed on Houji in ancient times. The Qin created forty-one counties, and Tai, Meiyang, Wugong were all individual counties.

> They were then combined into one. It had once been established [in the Tang] as Jizhou Prefecture, but soon was restored as a county. The soil of this terrain is fine and fertile, high and thick; its hills, highlands, knolls, and slopes stand grandly. The vegetation is rich; it is suitable for growing millet, soybean leaves, and soybeans.[32] The locals are good husbandmen, their customs are civilized and courteous, and all is appropriate to the area's glorious heritage in the Greater Ode.[33]
>
> 武功為甸內大縣，按其圖，古后稷封有斄之地。秦作四十一縣，斄、美陽、武功各異，至是合焉。蓋嘗為稷州，已而復縣。其土疆沃美高厚，有丘陵墳衍之大；其植物豐暢茂遂，有秬秠雈菽之宜。其人善樹蓺，其俗有禮讓，宜乎其《大雅》之遺烈焉.[34]

The author makes it clear in the very first sentence that a cartographic work was pivotally involved in the writing process. The key phrase "having consulted the map 案其圖" is the starting point of a detailed discussion of the historical changes in the county's political geography. It should be clear that the account of the historical changes was derived from the map the author consulted. This in turn means that the "map" offered both contemporary geographic information and a narrative account of the county's past. Therefore, it is likely that the "map" was actually a map-guide: map-guides were a kind of map-related geographic work that offered precisely such a temporal-spatial combination of knowledge, and as mentioned earlier, the map-guide of the county concerned would have been conveniently available to the author.

This passage sets the stage for Liu Zongyuan's political deployment of local geographic knowledge. It traces the origins of the county to Houji, the early legendary sage who was believed to have introduced farming to the Chinese people, thus igniting civilization in the central plains. The evocation of Houji is immediately followed by a skillful juxtaposition of the fertile fields, beautiful hills, and rich agricultural products of the county, creating a picture of material abundance in a blessed land. The triple repetition of the character "*da* 大," which variously signifies "grand," "majestic," and "magnificent," imparts a further sense of geographic grandeur to a county that had prospered throughout history. Here the map-guide, introduced through the phrase "having consulted," offers the cultural authority that the author needs to discuss and comment on the county's historical and cultural geography as an outsider. This sense of authority works nicely with the general tone of the passage, an official narrative that highlights the orderly prosperity of the local space. The map-guide's informed celebration of the land then provides the point of contrast between the past splendor of a historic county and its current state of dysfunction, which we will encounter in the following section of the piece.

The inscription then turns to a critical statement on how the region has been plagued by contemporary political turbulence, and the map-guide is again evoked with ambivalent personal feeling and political meaning. When Liu Zongyuan mentions the map-guide for the second time, he uses its image as an anchor for his mix of emotions about the current state of affairs in the local space and the moral censure he felt obliged to issue to those above. To illustrate the point, I quote the entire section here:

> On an unspecific day of the fifteenth year of the Zhenyuan Reign (799), the county changed its seat to Nanli. When the new town had been built, the walls where the old official inscriptions had been were damaged and the texts were missing. Yet no follow-up work was done. Three years later, Chen Nanzhong from Yingchuan took the post. The locals thought he was a good fit for the job, and referred to him as "simple and calm." So he sent his nephew Cun, who carried with him a map, to visit me and invite me to write a record. The territory and population of Wugong County are both so large that it is admittedly a difficult task for scholar Chen to govern it in a simple and calm manner. Emperor Gaodi of the Han Dynasty [Liu Bang (247–195 BC)] once announced to the world: for all those who were awarded ranks of nobility above the seventh and eighth categories, the local magistrates and vice magistrates had to respect them as their equals in an official manner. Therefore, it was very difficult for the local officials to do

> their jobs. Now the son of heaven is preoccupied with militarism and achievement, which is in a way similar to what happened in the beginning of the Han Dynasty. He distributed the imperial guards to protect the highways of counties, and Wugong County has the largest number of them. As Vice Magistrate in the region, Scholar Chen's duties include attending to local security, and his governing is still without failure. How can I possibly evaluate his merit? So I wrote this record.
>
> 貞元十五年某日，改邑於南里，既成新城，凡官署舊記，壁壞文逸，而未克繼之者。後三年，而潁川陳南仲居是官，邑人宜之，號為簡靖，因其族子存持地圖以來謁余為記。夫以武功疆理之大，人徒之多，而陳生以簡靖輔其理，斯固難矣。漢高帝嘗詔天下，凡以戰得首爵，七大夫公乘以上，令丞與抗禮，故為吏益難。今天子崇武念功，與漢初相類，分禁旅以守縣道，武功為多。陳生為丞於是，而又職盜賊，其為理無敗事，吾庸可度哉！為之記云。[35]

In the passage, Liu Zongyuan notes that the ideal unity of history, land, and culture in Wugong County that he has just outlined is endangered. The first sign of this suspension of local tradition is that a recent relocation of the county township destroyed all of the old office inscriptions. Because no previous office inscriptions were available to serve as potential models for Liu to write the new one, the vice magistrate had to send his nephew to take a map to him. The map is most likely the map-guide referred to in the previous section. Liu then turns to comment on the current state of the county: desirous of military achievement, the current emperor dispatched numerous imperial guards to defend transportation routes in the counties. Many of them ended up stationed in Wugong County. As the Qing Dynasty scholar He Zhuo 何焯 suggests, this is an implicit critique of the Dezong court, which allowed the imperial guards to harass civilians in the capital area.[36] In particular, in this sentence, the county's very name provides purchase for his criticism. By splitting Wugong into a phrase indicating the emperor's preoccupations with the military (*wu* 武) and achievements (*gong* 功), Liu sharpens his criticism of an emperor who has abandoned traditional ways and moral moderation.

Here, the information that the map-guide likely provides is part of what makes this veiled critique effective. Its presence forges a sharp comparison between the situations of the past and of the present. The map-guide, as the first segment earlier shows, attests to the peace and prosperity that once defined the land. The present, however, displays the degradation that occurred over time. The county magistrate was to be congratulated for his hard work in the hostile political environment, and the map-guide may have offered his only hope in his difficult years in office. Standing as a rare official document that had survived the county's chaotic displacement, the map-guide came to embody the much-desired continuity of local tradition, thus buttressing the political critique Liu intended to convey in the piece.

Like Yan Zhenqing before him, Liu Zongyuan used the geographical knowledge map-guides offered to craft a rich narrative about a local place and its precarious political trappings. The deliberate act of acknowledging the map-guide as a reference work further strengthened the political voice of the piece. Liu's strategic deployment of the geographic work empowered him to voice his political criticism of the imperial center from an entrenched local angle. Liu Zongyuan was not alone in using the empirical and symbolic power of local geography to deliver messages of greater significance. We shall now turn to Liu Yuxi, whose utilization of the map-guide was a bit more sophisticated, and no less fascinating.

Many scholars find that Liu Yuxi often engaged geographical sources in his writings about different localities, and that his texts describing local places are generally similar in style and structure to geographical prose.[37] In addition, as was noted in the Introduction, Liu is also known for having contributed several pieces to local geographical work while on local posts. In this sense, the following discussion of Liu's poem "Seventy Rhymes to Express My Thoughts in Liyang" and prose piece "Office Inscription of the Prefect of Hezhou" are but a small part of a much larger body of the author's geographically inspired literary work. Liu's literary engagement with the geographic genre speaks to the intimate relationship between the two in the mid-Tang local sphere and testifies to the presence of the geographical subtext in contemporary literary writing.

Liu Yuxi was transferred from Lianzhou to Hezhou to be its new prefect in 824. Shortly after his arrival in the prefecture, he composed a poem titled "Seventy Rhymes to Express My Thoughts in Liyang." The poem is meant to report to readers in the capital Liu's travel experiences between the two posts and his first impressions of Hezhou, as well as to express his personal thoughts and feelings. Relevant to our topic, the map-guide appears in the poem both as a critical source of local information and as a symbolic icon of the imperial authority which Liu represented in the local sphere. At the same time, Liu's artful reworking of local geographical information helped him direct his own political voice to his audience in the capital with added intellectual sophistication and persuasive power.

The map-guide makes an early first appearance in the preface to the poem:

> In August of the fourth year of the Changqing reign, my post was transferred from Kuizhou to Liyang, so I took a boat down the Min River, observed the Dongting Lake, passed Xiakou, reached Xunyang and turned east. My friend Cui Dunshi [Cui Qun] was dismissed from his position as a chief minister, and was serving as the governor of Wanling. He sent me a letter of invitation: "You must meet me before going to your prefecture. I will not let you go before we drink together for ten days." So I went to Wanling via Chizhou, as he wished. Dunshi later saw me off at the Jingting Shrine. From Gushu I crossed the Yangtze River and arrived in my prefecture. Since arriving here, I have consulted the map-guide, studied the present situation, and written poems about it, which await those who gather poetry and recite it at night.
>
> 長慶四年八月，余自夔州轉曆陽。浮岷江，觀洞庭，曆夏口，涉潯陽而東。友人崔敦詩罷丞相，鎮宛陵，緘書來抵曰："必我覿而之藩，不十日飲，不置子。"故余自池州道宛陵，如其素。敦詩出祖於敬亭祠下。由姑孰西渡江，乃吾圉也。至則考圖經，參見事，為之詩，俟采風之夜諷者。[38]

The preface outlines the author's long journey between the two posts. A quick list of midway destinations is followed by a brief statement about the author's first order of business in the new prefecture. He consulted the map-guide, studied the present situation of Hezhou, and expressed his thoughts in the form of poetry. In this busy but orderly narrative, the map-guide leads off the stately procession of a prefect's first acts in office, projecting a sense of power while showing the author's meticulous adherence to imperial expectations.

The map-guide arouses such associations because, as a centrally administrated but locally compiled geographical work, it both facilitates and symbolizes the imperial governance of the local sphere. The dual function of the map-guides as practical aid and symbolic token is elucidated in a poem by Zhang Ji 張籍 (767–830), a contemporary of Liu Yuxi, titled "Seeing Off Minister Zheng for Guangzhou" ("Song Zheng Shangshu fu Guangzhou" 送鄭尚書赴廣州):

> The sacred court selected the general to hold the credential tallies.
> When the imperial order is announced,
> it is heard by hundreds of officials.
> The barbarians north of the sea come to perform
> their greeting ritual,
> Districts and countries in Lingnan send in their map-guides.
>
> 聖朝選將持符節
> 內制宣時百辟聽
> 海北蠻夷來舞蹈
> 嶺南封管送圖經[39]

As the poem shows, the map-guide is the first source of comprehensive local information that an incoming official receives from his local subordinates. At the same time, at the inauguration ceremony for the new official, the map-guide, as a celebratory symbol, is juxtaposed with the imperial credential tally (*fujie* 符節) and the imperial order (*neizhi* 內制), two essential icons of the emperor's might within the four seas. An experienced official well versed in the political practices of the time,

Liu Yuxi did just what the official norms reflected in Zhang Ji's poem required of him, showing the care he applied to his official duties.

In addition to conforming to imperial norms, toward the end of the preface, Liu Yuxi also indicates his own political implications in utilizing the map-guide and writing the poem. Liu links the map-guide with poem collectors, practitioners of a long tradition of popular governance dating back to the pre-imperial period. In a process also associated with the compilation of the *Book of Songs*, legend has it that the kings of ancient times would dispatch poem collectors everywhere in their domain to travel deep into the countryside and collect folk poems. This was a pivotal way for a worthy king to stay in touch with the thoughts and desires of his people.[40] Seeing their times as marked by political dysfunction, Liu Yuxi and his reformist comrades like Bai Juyi took it upon themselves to revive this tradition and by so doing, to push the derailed train that was Tang politics back onto the right track. In their role as imperial governors of local populaces, they emphasized to the capital elite the political weight of the local sphere and the importance of revitalizing the dynasty's local constituencies. These ideals are made explicit in one of Bai's "New Music Bureau" (*xin yuefu* 新樂府) poems that eulogizes the poem-collecting tradition:

> The poem collector
> He gathers poems and listens to songs to guide the people's words.
> The singer has committed no crime and
>   the listener shall be forewarned.
> Down it flows and above it passes.
> Both those below and above are at peace.[41]
>
> 采詩官，采詩聽歌導人言
> 言者無罪聞者誡，下流上通上下泰[42]

In this reformist ideological context, the link Liu drew between the act of consulting a local map-guide and the practice of poem-collecting bore greater significance than adherence to the imperial laws. Here, reading the map-guide and studying the present conditions of local affairs were ways for a reformer to access the world of popular desires and local discontents, one step forward on the path to revitalizing the empire one locality at a time. It was at once a revelation of Liu's ambition and an idealistic injunction to his peers elsewhere in the realm.

The main body of Liu Yuxi's poem continues to engage the map-guide and the local geography it represents in various skillful ways. The poem contains seventy-four rhymes, totaling one hundred and forty-eight lines. To facilitate the discussion, I will first give a summary of the piece as a whole. The poem can largely be divided into four parts. The first part (line 1–line 16) draws directly from the map-guide and converts dry geographical knowledge into poetic lines with a pleasing literary quality, presenting an overarching picture of local culture, history and geography. The second part (line 17–line 72) recounts Liu's trip to Hezhou, especially his reunion with his long-time friend Cui Qun 崔群 (772–832) along the way. They spend days drinking together, a joyful and relaxed moment of relief from the traveling poet's onerous duties. As the poem develops, this hedonistic scene is quickly integrated into the narrative of a disciplined, aspiring official. The third part (line 73–line 124) returns to the local history and geography of Hezhou. The author assumes the solemn voice of the prefect, issuing both agenda-setting proclamations and pointed injunctions to the local officialdom. This is where the geographical information from the map-guide is again put to political use in an impressively eloquent literary manner. The fourth and final part (line 125–line 148) concludes the piece with an uplifting message to the court, displaying the author's political ideals while tying up the loose threads in the poem into a coherent statement about good governance and personal advancement.

The poem's intertextuality with the map-guide is concentrated in the first and third parts. The first part is an exemplary case of geographically informed information processing: the local knowledge found in the map-guide, most likely given in the form of rigid prose, is transformed into a refined poetic order bearing deep historical contemplations.

Once turned into a lake overnight,
This place has been a prefecture for a thousand years.
Here the hegemonic king lost his way.
Here the "second father" received his enfeoffment.
The Han established the position of Dongnan Wei.
The Liang Dynasty deployed an army close to the capital.
Once a violent Wu region,
Their dialect carried the vulgar accent of the Chu.
The "boiling well" has stopped flowing now.
The Wu River was once famous.
Laozi the imperial censor visited the mud platform.
Peng Keng took hermitage in a stone room.
Cao Cao's shrine still exits.
The Ruxu Fortress has not been leveled.
The tide swells with the moon.
The river rises with the spring.

一夕為湖地
千年列郡名
霸王迷路處
亞父所封城
漢置東南尉
梁分肘腋兵
本吳風俗剽
兼楚語音傖
沸井今無湧
烏江舊有名
土臺游柱史
石室隱彭鏗
曹操祠猶在
濡須塢未平
海潮隨月大
江水應春生[43]

What initiates the movement of the lines in this part is the ancient legend of Liyang being flooded into a lake following an earthquake. This story had already been evoked as an established and generalized metaphor of radical social disorder in many contemporary writings.[44] Nevertheless, Liu Yuxi must have also encountered it in the local map-guide, this time within the context of many other records of geographical objects specific to the Liyang region. The variegation of local landmarks and legends that the map-guide represents quickly finds literary variations in the first part of Liu's poem. In other words, unlike his contemporaries who drew upon the metaphor of the Liyang earthquake-flood out of its original geographical context, Liu inserts it back in the tangible local space, evoking it as the defining narrative of the place and the beginning moment of its history.

The first line of the poem, with its evocation of a natural disaster, is immediately counterbalanced by the place's long political history as a "prefecture," derived from the kind of knowledge that the map-guides offered in rich abundance. The historical weight is further enhanced by the revelation that it was in Liyang that the famous hegemonic king Xiang Yu lost his way while escaping the Han army and subsequently committed suicide. Coincidentally, Xiang Yu's chief councilor Fan Zeng 范增 (277–204 BCE), whom he respected as his uncle, or "second father" (*yafu* 亞父), received his enfeoffment in Liyang. In this couplet, the ingenious wit lies in the fact that Fan Zeng had advised Xiang Yu to dispatch the deceptive and ambitious Liu Bang 劉邦 (274–195 BCE) before Liu became a major threat; Xiang's neglect of Fan's advice eventually led to his defeat and Liu's coronation, changing the course of history.[45]

In the next four couplets, the poet moves on to further intrigue its learned readers. Deftly arranged allusions to famous or now obscure landmarks contrive to create a cultural mosaic, constituted by the histories of the Han Dynasty and Liang Dynasty, the customs of Wu, the accent of Chu, the "boiling well" and the Wu River, and two sites related to Daoist sages. The poet then invokes another dramatic coincidence to invite contemplation of history: the simultaneous presence of the temple for Cao Cao (曹操, 155–220) and the Ruxu Fortress on the local land. The fortress was built to defend the Wu Kingdom against the invasion of Cao, but its relics now stand side by side with the temple that worships its arch-enemy. Up to this point, the poet has created an intriguingly multilayered and well balanced historical experience associated with the local space via all the place names and allusions. The overview of the local landscape eventually culminates in nature's cyclical movement towards eternity: the tide swells in every lunar cycle and the river rises every spring.

This progression of the poem thus transcends the cultural landscape that it swiftly builds, assuring the reader that the poet is now speaking on a high level of dynastic histories and politics. Taken as a whole, these lines bespeak a capacious mind surveying the entire human past of a locality. It is not only a masterfully crafted report about the cultural geography of Liu's new territorial jurisdiction, but also a demonstration of the broad, balanced worldview appropriate to a capable poet-official.

The author does not repeat his explicit acknowledgement of the map-guide as a point of reference in this part; however, as the preface indicates, it was the source of the many names and events found in these succinct lines. The "boiling well," for example, was a rather obscure site recorded in the map-guide of Hezhou since ancient times. We know this because the *Records of Scenery on Earth* (*Yudi jisheng* 輿地紀勝), a geographic encyclopedia written during the Southern Song period, described the site as follows:

> Boiling well. According to the ancient map-guide, there used to be a boiling well in Hezhou. It was located in the ancient city a hundred steps from the western part of the county. During Emperor Yuan's reign in the Jin Dynasty, Guo Pu divined, "a well will boil in the county whose name has the character Yang." Soon after a well in Liyang began to boil. The location is unknown today.
>
> 沸井，古《圖經》云：和州舊有沸井，在郡西一百步古城內。晉元帝時，郭璞筮云："郡縣有陽名者，井當沸。"已而，歷陽縣中井沸。今失其處。[46]

Other geographical allusions in the poem were also likely drawn from the map-guide, or at the very least were readily available for reference in the map-guide. In this passage, Liu Yuxi displays extraordinary literary talent with this process of the poeticization of geographical knowledge: selecting a refined set of names and events from the map-guide, organizing them into verses that are thematically coherent and stylistically elegant, and arranging the lines in a way that manifests the poet's own view of political and cultural history.

The second part of the poem flashes back to the moment when Liu Yuxi learned of his reassignment and set out on the long journey. It combines typical travel landscape writing with a highly personal story about the author's encounter with his friend Cui Qun. As the later sections will show, this seemingly casual passage in an otherwise official poetic context subtly supports the author's political intentions and ambition. Before being demoted to Xuancheng, Cui was a long-time renowned official in the imperial court.[47] The lines carefully build an intimate relationship between Liu and Cui, which adds more weight to the author's political profile and therefore makes his persuasive arguments all the more convincing.

The third part of the poem marks the protagonist's entry into Hezhou. The traveler's casual observations are here succeeded by a rather self-congratulatory depiction of the welcome ceremony in Hezhou. With the tone reset to the proper official tenor, the author elaborates on local conditions in Liyang, shifting from the perspective of a political thinker to that of the official surveyor. The ancient legend of the earthquake-flood that opens the poem also finds its way into this section, reworked to serve new expressive purposes:

> There are wretched people in every house,
> And floods and droughts year after year.
> My deep thoughts often go to others instead of myself.
> All the orders must be announced beforehand.
> Faraway mountains line up into low screens,

Waterways wind like curving belts.
Fish in the lake are tastier than meat,
Wines from the official cellars are sweeter than candy.
I recall the past when the water source altered
And the earth's axis tilted in an instant.
Mount Hencoop turned into rocks,
And the eyes of the stone turtle on the county gate
    sank into the mud.
My responsibility is heavy because it concerns local life,
Although my rank is light according to public opinion.

比屋惸嫠輩
連年水旱幷
退思常後已
下令必先庚
遠岫低屏列
支流曲帶縈
湖魚香勝肉
官酒重於餳
憶昔泉源變
斯須地軸傾
雞籠爲石顆
龜眼入泥坑
事繫人風重
官從物論輕[48]

In the first couplet, Liu Yuxi's depiction starts with a bleak picture of the local situation, plagued by floods and droughts. The next couplet outlines the shapes of the faraway mountains and waters, but this interlude of landscape description soon dissolves into a laudatory depiction of the excellent quality of local fishing and farming products. Implied in the contrast between the poverty of local people and the abundance of local products is an implicit criticism of poor governance on the part of the presiding local officials.

Liu then brings back the historic earthquake-flood, completing the story with more legendary details associated with local landmarks. Together with the first part of the poem, this part alludes to the rich geographical subtext of the local map-guide, which likely contains the legends and with which it intertextualizes. At least one of the two legends, that of Hencoop Mountain, is recorded as having been included in the map-guide of Liyang.[49] Unlike the opening passage that provides an overarching picture of the local space, the geographically informed text here implies a clarion call for a specific locally based reformist politics. The two legends are embedded in the key couplet "Mount Hencoop turned into rocks, / And the eyes of the stone turtle on the county gate sank into the mud" (*Jilong wei shike, guiyan ru nikeng* 雞籠爲石顆，龜眼入泥坑). Both lines allude to folk legends about the Liyang earthquake-flood. The Mount Hencoop story tells how an old lady and her hencoop mysteriously turned into stones after witnessing the flood from atop a mountain. The myth of the turtle eye is more explicit in its moral teaching. A grateful prophet told an old lady that a flood would overwhelm her town if the eyes of the stone turtles in front of the prefectural mansion turned red. Frightened, she checked on the turtles every day, and eventually revealed the prophecy to a curious official. The careless official took it as a joke and painted the turtles' eyes red, perhaps just to tease the lady. It turned out he was teasing the heavens, and the heavens did not take it lightly. A mighty flood struck just as prophesied, and the old lady was the only one left to weep over the wreckage that was her hometown.[50] The myth is a forceful one, delivering an unmistakable warning to those who serve in the imperial realm. The people are the basis of imperial governance, and those to whom imperial power has been delegated must devote every fiber in their being to protecting their livelihood.

The teaching is certainly cliché just by itself, but Liu Yuxi was able to embed it in a locally rooted legend. More importantly, the poeticized story complemented Liu's local reformist political outlook, affording his political ideals a new layer of authenticity. The reader may be reminded of the tradition of poem collecting, which the author celebrated in the preface. The sage kings learned of public opinion by collecting folklore; the author might be doing just the same by collecting local legends from a respectable local geographic source. Reworking the two legends into a poetic couplet featuring a parallelism between a "Hencoop" (*jilong* 雞籠) and a "turtle eye" (*guiyan* 龜眼), the author uses the cultural authority the geographic subtext carried to show that his locally oriented political ideals are just as valid and authentic as the sage teachings in the classics. In other words, this couplet serves at once as a statement of the author's intense sense of duty and as a locally based justification for the political ideals he passionately pursued. The force of the line continues in the less equivocal line that immediately follows: "My responsibility is heavy because it concerns local life, / Although my rank is light in public opinion" (*Shi xi renfeng zhong, guan cong wulun qing* 事繫人風重，官從物論輕).

The fourth and concluding part of the poem gives an overview of the author's political career and crystalizes the political gist of the entire piece. In these final lines, a condensed numeration of his frustrations in the years of political exile leads readers to a final appeal, implicit but pressing, for the court to once again accept the author's offer of service:

My teeth have aged and medicine is a familiar thing.
My rank is low but I do not yield to dukes and ministers.
Those who hold up the sun are all grand senior men.
Those who spread the wind are all powerful governors.
It is time to summon the court-visiting officials
To pack up and go to the court for the new year.

齒衰親藥物
宦薄傲公卿
捧日皆元老
宣風盡大彭
好令朝集使
結束赴新正[51]

The author emphasizes his determination and ability to be of greater service to the majestic cause with a typical Tang rhetorical trick: even if physically sick and humble in rank, he shall stand firm and proud against the powerful and the prominent. Having linked himself rhetorically with those who serve his majesty ("hold up the sun") in court and those who spread his benevolence ("spread the wind") in the realm, he suggests it is time for the emperor to call him back to serve among the ranks of the worthy. The message is forged by way of an implicit metaphor. Instead of promoting himself, the author celebrates the court-visiting officials (*chaoji shi* 朝集使), who were local officials the court summoned to the capital for annual reports and inspection. The ceremonial scene of local officials entering the capital during the new year signifies the proper relationship between the court and its local constituencies. An imperial agent in the local sphere, the author implies through his invocation of the promising picture that such a relationship needs to be restored, like Liu's personal relationship with the court.

With this conclusion in mind, we are in a better position to make sense of the subtlety and inferences in the earlier parts. The author is reporting local conditions back to the court, but he is also constructing a new image of himself as a proud official serving the emperor in the local sphere. He follows the official norms in the preface, displays the right amount of moderation and intellectual breadth in the grand narrative about the human vicissitudes of Liyang, and then implies his rapport with Cui, a widely respected politician, through a recounting of their happy reunion on Liu's way to Hezhou. In this political reading of the poem, the carefully summoned images from the local geographic source empower the author, both in his routine governance as a prefect and in his conversations with the political elites in the capital.

Interestingly, this long poem was originally inscribed on the east wall of Liu Yuxi's local office, while on the facing wall in the opposite direction was an office inscription also by Liu.[52] These two texts form a unique binary star system, both in this official space and in Liu's literary collected works. Pertinent to our discussion about the cross-field multilayering of mid-Tang local writing is the intertextuality between the poem and its twin prose, which confirms that the map-guide was also underlying the latter. The following is the passage that reminds us of Liu's poem:

> The stronghold is built on Mount Liang, and the water is called Lake Li. Agricultural products include four major grains and the livestock include all six major animals. The cellar keeps delicious wine and the kitchen holds juicy fish. Stories about the immortals are circulated among the people. Yuanyuan had his terrace, and Peng Keng had his cave. There is a famous mountain named the Hencoop and a famous fortress named Ruxu. The occult locations include the bloody threshold and the auspicious locations include the boiling well. The town walls are high and sturdy, and were built by Fan Zeng.
>
> 鎮曰梁山，浸曰歷湖。田藝四谷，豢全六擾。廬有旨酒，庖有腴魚。神仙故事，在郊在藪。玄元有臺，彭鏗有洞。名山曰雞籠，名塢曰濡須。異有血閫，祥有沸井。城高而堅，亞父所營。[53]

This passage is a variation of Liu's earlier poetic work. The line "the cellar keeps delicious wine and the kitchen holds juicy fish" (*Lu you zhijiu, pao you yuyu* 庐有旨酒，庖有腴鱼) paraphrases the couplet "Fish in the lake are tastier than meat, / Wines from the official cellars are sweeter than candy" (*Huyu xiangsheng rou, guanjiu zhongyu xing* 湖魚香勝肉，官酒重於餳). "Yuanyuan has his terrace and Peng Keng has his cave" (*Yuanyuan youtai, Peng Keng youdong* 元元有台，彭铿有洞) is a rewriting of the couplet "Laozi the imperial censor visited the mud platform. / Peng Keng took hermitage in a stone room" (*Tutai you Zhushi, shishi yin Peng Keng* 土台游柱史, 石室隱彭鏗). Mount Hencoop, the Fortress Ruxu, the bloody threshold, the boiling well and the sturdy town walls built by Fan Zeng are all featured in the poem in its various constituent parts. Although different in genre and style, the two pieces of local literature devoted to Hezhou clearly shared the same source of geographic information.

To summarize the discussion in the chapter up to this point, from Yan Zhenqing to Liu Zongyuan and onto Liu Yuxi, map-guides, poetic creation, and official writing were deeply entangled in a highly politicized local sphere. They worked in concert to generate the sense of authority and cultural authenticity that authors needed in their efforts to join empire-wide political conversations from their often-neglected local spaces. The trio blended into one literary voice, giving these writers greater political expressivity as they strove to reform the empire.

MAP-GUIDES, THE LITERATI, AND THE GEOGRAPHY OF FOLK CULTS

As my discussion of Yan Zhenqing's inscriptions has shown, map-guides served as the spiritual map of a given locality, rendering the otherwise hazy world of folk cults legible to agents of the Tang state. In this regard, the Tang map-guides continued to embody the spiritual dimension that is commonly seen in Chinese geographical work ever since the ancient *Classics of Mountains and Seas.* They mediated between the human and the divine, the imperial and the local. While the court did issue general directives on the handling of local cults, it was often the locally compiled map-guides that allowed the literati-officials to fill in the gap between the blanket imperial guidelines and the complex conditions on the ground. Court-appointed local officials relied on the map-guides to locate, inspect, and evaluate local religious sites. In addition, Tang map-guides also intervened in the contested process as an active medium between the local community and the officialdom. Instrumental in these processes was a special literary genre of the local sphere, namely the orisons of appeal. My analysis in the following pages will show that this genre of official writing, often overlooked by modern literary scholars as formulaic and pragmatic, was actually a dynamic textual space in which talented authors reorganized geographical information from the map-guides into forceful yet subtle political utterances.

Unlike Daoism, which the court sponsored throughout, the Tang state viewed local folk cults with apprehension; nevertheless, it had never been able to exert full control over the matter from the center, but rather had to rely heavily on local governments.[54] According to the historian Lei Wen, throughout the dynasty, the court issued several statements prohibiting illegitimate shrines (*yinci* 淫祠)—generally denoting those worshiping un-Confucian historical figures and demonic spirits—but it remained unclear in these statements how local officials were to judge specific local deities as such.[55] The court also provided general ritual codes such as the *Kaiyuan Ritual Code of the Great Tang* (*Da Tang Kaiyuan li*, 大唐開元禮, completed in 732) on proper local worship, but these codes were never fully implemented across the empire.[56] After the Rebellion, destabilizing local political conditions gave the court further incentive to delegate the power to regulate the local spiritual sphere to its agents in the field.[57]

As prefects and magistrates found themselves in a world of obscure local gods and baffling shrines, the map-guides came in handy as a convenient source of guidance.[58] In the map-guides, one could find all kinds of information about local shrines, including their physical locations, symbols, times of sacrifice, founding dates (if available), and sometimes their putative functions and efficacy. In addition, some map-guides also contained instructions for officials and locals on how to interact properly with the gods at folk shrines. In other words, local map-guides put the various local deities in a legible imperial order and defined for the reader their nature, responsibility and ritual requirements.

Map-guides, then, became important collations of information about shrines and their associated gods. The previous discussion of Yan Zhenqing included a map-guide record of the Master of Rain, and the following are several other examples of a similar kind from the same map-guide:

> The Earth God:
> The above is one *li* to the south of the prefectural city. A shrine was established to house an icon of the god. People pray to this god when there are disasters and unrest in the territory. The founding date of the shrine is unknown.
>
> 土地神

右，在州南一里。立舍，畫神主。境內有災患不安，因以祈焉。不知起在何代。

The Earl of Wind:
The above is fifty paces northwest of the prefectural city. A shrine was established to an icon of the god. People pray to this god when there are calamities in the territory. The founding date is unknown.

風伯
右，在州西北五十步。立舍，畫神主。境內風不調，因以祈焉。不知起在何代。

The Zoroastrian God:
The above is one *li* east of the prefectural city. A shrine was established to house an icon of the god. There are twenty niches all told. The yard is one hundred paces in circumference.

祆神
右，在州東一里。立舍，畫神主。總有廿龕。其院周回一百步。[59]

These succinct entries provide to readers the basic information they would need to find and make sense of the shrines concerned. Other entries are more sophisticated, informing the reader not only of matters of basic nomenclature but also of the locally circulating legends and ritual practices associated with the deities:

> The villagers prayed to her each April. High-ranking prefectural and county officials had to pay obeisance to her before entering the shrine. There suddenly appeared a dark stone in front of the shrine. Three feet in length and width, and two feet in height, it had an old inscription, "This is the stepping-stone for the lady's husband to mount his horse." It remains there to this day.
>
> 鄉人每到四月祈禱，州縣長吏若謁此祠，先拜然後得入。於祠前忽生青石一所，縱橫可三尺餘，高二尺餘，有舊題云："此是姑夫上馬石"。至今存焉。[60]

Writer-officials utilized such information from the map-guides in various ways to determine and then communicate their agendas of local spiritual governance to their superiors. The Daozhou prefect Yuan Jie, for example, wrote an appeal for imperial support for the local worship of Shun around its historicity as manifested in the map-guide. Next is the text from his memorial to the throne, justifying the tradition and the administrative concessions necessary to revive it:

> According to my careful consultation of a map of the land, Shun's tomb was on Jiuyi Mountain, and Shun's temple was by Taiyang Creek. The tomb has been lost to history; there remains no trace of Taiyang Creek. In the Qin and Han periods, a temple has been located at the foot of the mountain. After many ages, all traces of the temple are gone. Every time there was an imperial command to perform sacrifices in the prefecture and the county, officials could only carry out the command in the wilderness. Shun had great virtue and created a grand enterprise and was a role model for hundreds of kings. How could he die in the desolate land without his own tomb and temple? I, your subject, respectfully followed the conventional practice and had a temple built on the West Mountain. Here I particularly entreat your heavenly grace to allow the waiver of tax on a few households near the temple, and order that they seasonally sweep and clean it, as a constant model. How would this only display the sage's perfect virtue down through countless generations? Truly I wish to manifest your majesty's fundamental beneficence inexhaustibly.

> 謹按地圖，舜陵在九疑之山，舜廟在太陽之溪。舜陵古老以失，太陽溪今不知處。秦漢已來，置廟山下，年代寖遠，祠宇不存。每有詔書令州縣致祭奠酹，荒野恭命而已。豈有盛德大業，百王師表，歿於荒裔，陵廟皆無？臣謹遵舊制，於州西山上已立廟訖。特乞天恩，許蠲免近廟一兩家，令歲時拂灑，示為恒式。豈獨表聖人至德及於萬代？實欲彰陛下玄澤及於無窮。[61]

The author refers at the very beginning to "a map of the land" as his source of knowledge about the Shun Shrine. As the passage goes on, we learn that this "map of the land" is more than a simple map: it does not really show the present location of the shrine, as "the tomb has been lost to history; there remains no trace of Taiyang Creek." Rather, it contains textual statements about the tomb's history and past location in relation to mysterious geographic objects long lost to the passage of time. In other words, the so-called "map of the land" is in fact a text about local geography, most likely a local map-guide. The text delineated for the author the remote history of the shrine, which the author then juxtaposes with the recent desolation of the iconic site. Here the map-guide emerges as a container of an (imagined) golden past in textual form while offering the reader a tacitly contrasting perspective. Only then does the author present his argument for imperial benevolence, which appears both legitimate and pressing in the deftly constructed geographic context.

Writer-officials may have also used map-guides to locate shrines deviating from the Confucian orthodoxy. In these cases, map-guides presented traces of alternative traditions, and allegiances exposed themselves to the official eye. Liu Zongyuan's inscription about the destruction of the temple for the Bi Pavilion god, also composed in Daozhou, is a case in point. The inscription, titled "Record of Destroying the Bi Pavilion God in Daozhou" ("Daozhou hui Biting shen ji" 道州毀鼻亭神記), narrates how the author's friend and Daozhou prefect Xue Bogao 薛伯高 (fl.810s) finds out about a local shrine to the wicked figure Xiang "after investigating local customs and reading the map of the land" (*kao minfeng, pi ditu* 考民風，披地圖).[62] Xue was infuriated by the discovery, because ironically, Xiang was the brother of Shun and allegedly had attempted to kill Shun. Outraged by the worship of this notorious fratricide condemned in the Confucian tradition, Xue ordered "the shrine removed, the land turned to ruins, and the idol dropped deep into the river" (*chen qiwu, xu qidi, chen qizhu yu jiang* 撤其屋，墟其地，沉其主於江).[63] In this interesting case, the map was likely part of the local map-guide that was readily available to the prefect, and the text of the map-guide likely celebrated a site of worship that failed to comply with imperial cultural norms. The prefect was acutely aware of the intermediate nature of the genre of map-guides and, taking cues from its unwitting eulogy, exerted his court-endowed power in the local spiritual space.

However, map-guides were not always passive conduits of knowledge awaiting official judgment. The geographic texts could contain in themselves a persuasive agenda that might at times have intervened, however subtly, in readers' perceptions of local traditions. In the *Map-Guide of Shezhou* (*Shezhou tujing* 歙州圖經), compiled during the mid-Tang, for example, we find a fantastic story that may have justified an otherwise heterodox local god. The story tells how the historical figure Jin Shang 靳尚 (?–311 B.C.E.), who had enjoyed local worship for many centuries, converted to Buddhism thanks to the enlightenment of the monk Fadu 法度. Jin Shang had long been considered an envious enemy of the loyalist poet Qu Yuan 屈原 (c. 343–278 B.C.E.).[64] Therefore, shrines for him were conventionally considered illegitimate. This can be seen from the poem "Minister of Documents Li of Zhexi Reported to the Court His Abolishment of Illegitimate Shrines" ("Zhexi Li Shangshu hui Yinhun miao" 浙西李尚書奏毀淫昏廟) by the mid-Tang poet Xu Ning 徐凝 (fl. 810s): "We have heard that illegitimate sacrifices were abolished, / The mountains for miles around shall fall quiet. / To ease Lingjun's [Qu Yuan] remorse, / Jin Shang's shrine must be burnt first" (*Chuanwen fei yinci, wanli jing shanbei. Yuwei Lingjun hen, xianshao Jin Shang ci* 傳聞廢淫祠，萬里靜山陂。欲慰靈均恨，先燒靳尚祠).[65] The poem makes it clear that the temples for Jin Shang were always the first targets to go down in anti-unorthodoxy movements. In the version of Jin Shang's biography in the map-guide, however, Jin Shang does the right thing by converting wholeheartedly to Buddhism, a more institutionalized mainstream religion:

> [Jin Shang] saluted and said, "I, your disciple, have owned this mountain for more than seven hundred years. . . .You, the master, are the embodiment of the way and the power. Now I wish to offer myself in service to you and be ordained with the Five Precepts, and bind our fate together forever." Du says, "The ways of humans and the ways of gods are separate, and you cannot force me to do

> this. Moreover, you, the almsgiver, have enjoyed blood sacrifice for ages, and this is especially forbidden by the Five Precepts." Jin Shang said, "If I can become your disciple, then I will first put an end to the killing." Then he left. The next day someone sent Du ten thousand coins, as well as incense and candles and so on, with a note saying, "Offered by the disciple Jin Shang." On the fifteenth day of the month, Du organized a ritual. Shang came again. He circled with temple with the others and took part in the ritual, then left after his ordination. Soon the shaman of the shrine on She Mountain dreamed that the god [Jin Shang] told him: "I have already been ordained by Du, the Dharma master. From now on do not kill for the sacrifice." Since then the shrine only offered the god rice and vegetables.
>
> (靳尚) 致敬畢，乃言："弟子主有此山，七百餘年矣。....法師道德所歸，謹舍以奉給，並願受五戒，永結來緣。"度曰："人神道殊，無容相屈，且檀越血食世祀，此最五戒所禁。"尚曰："若備門徒，輒先去殺。"於是辭去。明旦，一人送錢一萬，並香燭等，疏云："弟子靳尚奉供。"至其月十五日，度爲設會，尚又來，同衆禮拜行道，受戒而去。既而攝山廟巫夢神告曰："吾已受戒于度法師矣，今後祠祭勿得殺戮。"由是廟中薦獻菜飯而已。[66]

In the story, Jin Shang followed Fadu's teaching to stop ritual animal sacrifice at his shrine and was thus initiated into the monkhood. The story can be read as a Buddhist conquest of a folk cult, but it can also be read as the rehabilitation of the local god Jin Shang. At a time when local map-guides were widely consulted in the regulation of local shrines, a story like this might have helped to protect the shrine to Jin Shang from state suppression.

In addition to helping local officials access and cleanse the local spiritual space, map-guides also facilitated their ritualistic performances as dutiful agents of the majestic throne. Beginning in the High Tang period, regional governors often put paying tribute to proper and prestigious local shrines as the first item on their lists of official duties upon arrival at a new post.[67] In 819, for instance, Han Yu composed a prayer text for the local god of the great lake, so that a surrogate could read it aloud at the shrine in his absence. Local officials were also expected to pray to local gods for the safety and wellbeing of the local population. For the mid-Tang poet-officials, then, the performance of proper rituals at local shrines was not only their official duty, but also a measure of their work ethic and worthiness. As Li Deyu says in "Essay on Supplications" 禱祝論, "When there is an untimely lack of rain and the crops are on the verge of withering away, if the governor criticizes himself within his closed-off hall, the common people will not see it. If he does not make a circuit around to all the various sacrificial sites, then they will all say that the governor shows no solicitude toward the people, and their laments will be heard unceasingly as far away as the farm fields" (*Shishi buyu, jiase jiangku, bige zegong, baixing bujian. Ruofei bianzou qunwang, ze jiewei Taishou wu youren zhiyi, suizai quanmu, bujue tanyin* 失時不雨，稼穡將枯，閉閤責躬，百姓不見。若非遍走群望，則皆謂太守無憂人之意，雖在畎畝，不絕歎音).[68]

These ritualistic performances gave rise to a special genre of literary writing known as orisons of appeal. Written by the presiding official of a local place in imperative mode, they were addressed directly to the officially recognized local deities and read aloud at their shrines. Famous writers like Han Yu, Yuan Zhen, Liu Zongyuan, and Bai Juyi all engaged in such writings in their official capacity, usually in the hope that their prayers would bring benign weather and spare the locality from natural disasters. Their prose pieces were dedicated to a variety of local deities from the well god, dragon god, and earth god to the lake god, Shun god, and Lord Huang mentioned in chapter 1. For example, in Liu Zongyuan's orison for the Shun god in Yongzhou, titled "Appeal for Sunshine at the Shrine for Shun" ("Shunmiao qiqing wen" 舜廟祈晴文), the author asks Shun to halt the ferocious thunderstorms and send sunshine to the land.[69] Bai Juyi also wrote a piece begging the King Qiu God (*Qiuwang shen* 仇王神) to restrain a tiger that had eaten several men.

The orisons of appeal are noteworthy for our discussion of literary geography because authors often distilled information about the local spiritual sphere from the map-guides and, in a tacit demonstration of their literary talent, developed such information into archaic prose pieces displaying rhetorical power. That is to say, under their shared formulaic façade, these essays were in fact locally specific literary works built upon the subtext of local geography. The interaction between the geographic source and the official literary genre again bespeaks the productive synergy between geography and literature in the local sphere. To illustrate these points are two orisons of appeal, one by Yuan Zhen in Tongzhou, the other by Bai Juyi, composed in 823, when Bai was the prefect of Hangzhou. Yuan's orison, titled "Informing the God of Shezhu Mountain" ("Gao Shezhu shanshen wen" 告畬竹山神文), appeals to the local mountain god to approve a reclamation of land in the god's domain. The following passage shows explicitly that the author consulted local geographic sources and conceived of the prose with the specific local context in mind:

> According to the records of Tongzhou, the mountains and rivers were named after the god, and your mountain is close to the houses in town and supports the road. Yet for almost fifty years there has been neither ploughing and harvest, nor pillars and columns [on the mountain], since cheetahs and serpents have gathered to cover and destroy the paths. Indeed this is due to the fact that human power cannot manage the mountain, and it is not the god's fault.
>
> 按通之載，號神為名山川，且邇邑屋而扶道途，然而不砟不穫，不礎不柱，蘩集貙蟒，蔽弊道路，將五十年矣，實人力不足以於山也，非神之過也。[70]

The author refers to the "records of Tongzhou" as the basis of his appeal. Since the "records" contained the name and nature of the local deity as well as local geographic conditions surrounding the mountain, it is possible that the local source was the local map-guide. In fact, most orisons of appeal do not make their geographical references explicit, but from Yuan's case we know that the local map-guides were likely the source of information about the deity in question. With the help of the geographical reference, local officials were able to arrange the rituals according to local customs and open an informed dialogue with the gods in their orisons of appeal.

In light of this probable assumption, we can now turn to Bai Juyi's "Sacrifice to the Mount Gaoting God" ("Qi Gaoting shen wen" 祈皋亭神文), which appealed to the Mount Gaoting God for rain:

> In the third year of Changqing reign [the year of guimao], the seventh month [the month of guichou], the sixteenth day [the day of wuchen], I, Bai Juyi, Grand Master for Court Discussion, commissioned with extraordinary powers over the military affairs of Hangzhou, concurrently prefect of the said prefecture, Supreme Pillar of the State, offer wine, milk, incense, and fruits to you the god of the Temple of Gaoting, and state to you my solemn wish. The heat was overwhelming last fall, and rain is scarce this summer. I worry deeply about the disaster that is besieging the people of Hangzhou. In humbleness I received imperial commission; with shame I admit my lack of competence. Having witnessed the draught, I dare not rest unperturbed. Over the past days I have prayed to the divine Minister Wu, and pleaded in the temple of the city god. The gods responded to my plea, yet rain did not moisten my feet. Thus I have selected this date for the divine affair of pleading instead to you the god. I have humbly learned the following about you the sagacious god: you derived your spiritual powers from the god of the earth, and acquired your goodness from the Buddha; you are keen-eared, clear-sighted and righteous; virtuous, peaceful, and benevolent. There are no spirits with which you cannot connect; there are no prayers to which you do not respond. Now please allow me to pray to you in sincerity; please examine my request in its totality. If within the four borders, during the next five days, the rain is plentiful and the crops are moistened, I shall expand your divine portraits, present you with fragrant incense, organize orchestral performances, and offer treasures to repay the favor. Such will not only be a blessing to us humans, but also a glory for you the god. Should you remain in

silent solitude, unconcerned and unresponsive, refuse to answer the pious plea of the head official, fail to recognize the earnest wishes of the petty people, sit unmoved as you view the peasants, their farms remaining arid—if this be the case, it would not only be a plight for us humans, but also an embarrassment for you the god. Dear god, please choose! I shall wait for your judgment in reverence. Please enjoy this sacrifice.

維長慶三年歲次癸卯，七月癸丑朔，十六日戊辰，朝議大大、使待節杭州諸軍事、守杭州刺史、上柱國白居易，以酒乳香果昭告于皐亭廟神：去秋愆陽，今夏少雨，實憂災沴，重困杭人。居易忝奉詔條，愧無政術；既逢愆序，不敢寧居。一昨禱伍相神，祈城隍祠，靈雖應期，雨未霑足。是用撰日祗事，改請于神。恭聞明神，稟靈於陰祇，資善於釋氏。聰明正直，潔靖慈仁。無幽不通，有感必應。今請齋心虔告，神其鑑之。若四封之間，五日之內，雨澤霈足，稼穡滋稔，敢不增修像設，重薦馨香，歌舞鍾鼓，備物以報？如此，則不獨人之福，亦惟神之光。若寂寥自居，肸蠁無應，長吏虔誠而不答，下民顒望而不知.坐觀田農，使至枯悴。如此，則不獨人之困， 亦唯神之羞。惟神裁之，敬以俟命。尚饗。[71]

This orison of appeal demonstrates Bai's familiarity with the minute details of the local spiritual sphere in Hangzhou, where he served as the prefect. He knew precisely which local deities to turn to in case of drought and worked his way carefully up the ladder of divine hierarchy. Before pleading to the god of Mount Gaoting, he had already prayed in the Minister Wu shrine and in the temple for the city god. The Minister Wu shrines were devoted to Wu Zixu, a worthy statesman of the Spring and Autumn era. He was deified posthumously in a few lower Yangtze prefectures that benefited from his capable governance in turbulent times. The city god, on the other hand, was a commonly worshiped local god in many places. This was not a legitimate deity sanctioned by the central court; rather, the central court prescribed that it was up to the local government to determine the ritual code for it. To learn about these places of worship and to select the right sites to perform the rain ritual,

Bai would have had to turn to sources of information in the local sphere. In addition, the phrase "I have humbly learned the following about you the worthy god" (*gongwen mingshen* 恭聞明神) also suggests that Bai studied local sources diligently before deciding his course of action. Since it was a common practice for prefects of the mid-Tang and onwards to consult local map-guides upon their arrival in a new post, and map-guides contained precisely these items of information on local cultural geography, the map-guide was likely among the sources Bai consulted, very much as Yuan Zhen did in Tongzhou.

In Bai's orison, his precise understanding of the local deities is matched by the literary power of the elegant prose. Bai builds the piece forcefully, establishing rhetorical power through stylistic arrangements. He creates a direct conversation with the silent mountain god, appealing to a reciprocal and collaborative relationship. His modest self-admonition and unreserved acclamation for the local deity form a neatly contrasting parallel, at once conforming to ritualistic norms and exerting moral pressure on the imagined divine audience. Bai then expresses his expectations of the god, and follows with the warning that should the god not respond to his request, it will not only mean trouble for the local people, but will also shame the god. Here, the writer treats the god as his equal, pressing it with both threat and promise, blame and reward. The prose betrays a sense of urgency, making the author's anxious desire for local wellbeing vivid and tangible.

These examples from Yuan Zhen and Bai Juyi show that works in this genre had a definite literary dimension. Granted, the Tang court did issue sample texts to guide the composition of orisons in the *Kaiyuan Ritual Code of the Great Tang*, and these texts became the models for local officials' own orisons of appeal.[72] However, writers did not simply withhold their creative vigor to comply with these restrictive norms. Despite the officially stipulated formulaic structure of Yuan and Bai's orisons, these pieces are rich in rhetorical tricks specifically designed to add to their power of persuasion. They cared as much about the literary quality of their compositions as their efficacy in the spiritual realm. For instance, the mid-Tang writer Liu Ke 劉軻 (fl. 819) mentioned in his "Farmer's Supplication" ("Nongfu dao" 農夫禱) that he heard a farmer praying in a local shrine in simple language. Sympathetic to the farmer, Liu rushed up to help polish the text for him.[73] This is an explicit testament to the mid-Tang writers' unyielding pursuit of literary aesthetics, even when constrained by the highly regulated form of the genre.

In addition to their literary quality, some orisons of appeal also bore specific political implications. Alexei Ditter observes that in the eighth and ninth century, orisons for appeal were not only performed in shrines, but also circulated within the literary community. He incisively points out that "recognition by this larger audience as skilled in ritual matters could help the advancement of one's career."[74] In this sense, the literary quality found in these formulaic essays is in itself a political performance. Ditter goes on to argue that skilled writers used orisons of appeal to speak surreptitiously to the different audiences in the literary community.[75] He quotes Han Yu's "Sacrifice to the God of the Great Lake" ("Ji Dahu shen wen" 祭大湖神文) as an example. In it, Han requested that the lake god end the ceaseless downpour, because if it continued, "the farmers and mulberry pickers would be without the means by which to pay their taxes and tribute" (*nongfu sangfu, jiang wuyi ying fushui* 農夫桑婦，將無以應賦稅). Ditter proposes that "Han's emphasis on the people's inability to pay their 'taxes and tribute' seems designed as much to convey to his superiors in the central government the fiscal ramifications of the flooding as to express to these deities the urgency to do something about the weather."[76]

For these reasons, I consider orisons of appeal a literary genre that manifests the mid-Tang writer-officials' locally based geographical understanding, their creative skills in producing aesthetically accomplished prose within the norms of the officialdom, and, in some cases, their strong but oblique political criticism. Like office inscriptions and report poems, they are a map-guide-informed set of writings composed in the local sphere, a fine product of the cross-field exchange between literature and geography. In this section, we have seen that local geographical works gave local spiritual traditions a spatial order and subtly manipulated official regulation through carefully considered stories of local deities; other records and memorials concerning local deities drew upon the map-guides to form arguments about local governance. Orisons of appeal, embodying the most performative component of a local official's engagement with the spiritual dimension of a given local sphere, further extended the empire's power into that local sphere. In this sense, local geographic works like map-guides figured prominently in the intermediate space between the empire and its local constituencies, while at the same time enabling writer-officials to play the role of mediator between the two as imperial agents.

Map-Guides and Travelling Literati

Moving away from the forms of writing with official purposes that we have examined so far, in the last section I discuss travel poems that engage with map-guides. Poems like Han Yu's "On Arriving in Shaozhou, I Sent this Poem to Prefect Zhang, the Former Attendant Censor, to Borrow a Copy of the Map-Guide" ("Jiangzhi Shaozhou xianji Zhang Duangong shijun jie tujing" 將至韶州先寄張端公使君借圖經) (819–820) and Wang Jian's "On the Tombstone of the Interior Official Cai in Suanzao County" ("Ti Suanzao xian Cai Zhonglang bei" 題酸棗縣蔡中郎碑) show that the map-guide was used as a sightseeing reference for traveling literati. They testify to the popularity of the map-guides among the literati on their non-official trips. In addition, the explicit evocation of map-guides in both cases reveals the author's complicated feelings regarding personal memories and historical changes.

Let us first consider Han Yu's poem:

> I have long heard of the mountains and rivers of Qujiang,[77]
> But feared it would be doubly hard to visit not knowing

their names.
I would like to borrow the map-guide as I cross the border,
So I may open it each time I come to a beautiful place.

曲江山水聞來久
恐不知名訪倍難
願借圖經將入界
每逢佳處便開看[78]

At the time, Han was about to pass by Shaozhou in the Lingnan area as part of another of his political relocations. As the title of the poem shows, before entering the territory of Shaozhou, he wrote the poem to his old friend, the Prefect of Shaozhou, asking to borrow the local map-guide. Han's desire to borrow the map-guide indicates that he was determined to go on a detailed and informed sightseeing tour in Shaozhou. It is not surprising that Han would consider borrowing a map-guide to facilitate his tour, given his intimate relationship with maps in the context of his professional career. Rather, the question for us to investigate is how the episode of borrowing a map-guide functions in the poem. The poem lacks the aesthetic verve and linguistic novelty that distinguish much of Han's poetry, but a contextualized reading of it reveals that it holds more import for the author than a merely "occasional" poem would.

The first couplet of the poem, "I have long heard of the mountains and rivers of Qujiang, / But feared it would be doubly hard to visit not knowing their names" indicates Han's lack of experience with the landscape in Qujiang. But Han was by no means entirely unfamiliar with the area. First of all, Shaozhou held an important place in Han's childhood memories. At the age of twelve, Han Yu travelled to Qujiang in Shaozhou as a dependent of his older brother Han Hui 韓會 (738–779), who had been demoted there by the court. Han Yu stayed in Shaozhou for two years before Han Hui died at the age of forty-two.[79] For people in the Tang, Qujiang in Shaozhou connoted a place close to the empire's southern edge. For Han Yu, it was the destination at the end of a rough journey that exposed him for the first time to diverse landscapes. In Han's autobiographical writing, Qujiang would become a place that evoked the turbulence in his early life more than a particular topography. In his autobiographical essay "Rhapsody on Recovering My Purpose" ("Fuzhi fu" 復志賦), he recalls his experience as follows:

> Not yet twelve years old, I followed my brother and moved to the south. We confronted the startling waves of the Yangtze River, and crossed the boundless Dongting Lake. We rested only upon arriving in Qujiang, after climbing over the mountain range at the empire's southern limits. I sighed at the quick passing of the years, and took my orphaned nephew and widowed sister-in-law back to the north. Then troubles happened in the central plain, and we moved south of the Yangtze River to make a living.
>
> 當歲行之未復兮，從伯氏以南遷。淩大江之驚波兮，過洞庭之漫漫。至曲江而乃息兮，逾南紀之連山。嗟日月其幾何兮，攜孤嫠而北旋。值中原之有事兮，將就食於江之南。[80]

This passage conjures up an image of a young child who is overwhelmed by his remote trip to the south, the formidable terrain along the way, and the successive misfortunes that befell his family in the south.

The next time he encountered Shaozhou, his situation was little better. Having been demoted, he was on his way from the capital to Chaozhou, a place in Lingnan even further from the capital than Shaozhou. On his way to Chaozhou, Han lost one of his daughters; he refused to bring his family members to the desolate Chaozhou and kept them all in Shaozhou.[81] While passing Shaozhou, new and old adversities intertwined in his mind, and he wrote the poem titled "Expressing My Thoughts Passing the Mouth of the River in Shixing" ("Guo Shixing jiangkou ganhuai" 過始興江口感懷):

> Remembering I followed my older brother here as a child,
> Now coming south again I am the sole survivor.
> Before my eyes a hundred mouths chase one another,
> But there is no one to talk with about the past.

憶作兒童隨伯氏
南來今只一身存
目前百口還相逐
舊事無人可共論[82]

In Shaozhou, Han found himself immersed in deep grief for the brother who had brought him to Shaozhou for the first time many years before. Although accompanied by many family members and attendants, the adult Han finds that he has no one to talk to about his memories. The busy present contrasts with the painful southern experience of his childhood, pushing him into a deep, lonely melancholy.

In contrast, Han's trip to Shaozhou in 819 or 820, the trip for which he decided to borrow the map-guide and arrange a sightseeing tour, was celebratory.[83] This time, Shaozhou was the midpoint on his way to a regional promotion from Chaozhou to Yuanzhou. Yuanzhou, as an administrative district, was one level higher than Chaozhou, and was a much better place overall. Furthermore, when Han was demoted to Chaozhou, he was committed to five years of service there. In the end, however, he received amnesty and a promotion after merely seven months of service. Han was understandably delighted. There were also other reasons that made this Shaozhou trip pleasant. Since Han had left all his family members in Shaozhou, a family reunion awaited him there. The time he spent in Shaozhou also happened to overlap with the Chinese New Year. Finally, the person from whom Han requested the map-guide, surnamed Zhang, was Han's close friend who had always been concerned about Han's political misfortune. He took care of Han's family in Shaozhou and wrote a congratulatory poem upon hearing that Han would be relocated from Chaozhou to Yuanzhou.

As is clear, before requesting to borrow a map-guide, Han had experienced the Qujiang area several times: in his childhood, again when he was relocated to Chaozhou, and likely on other occasions when he lived in Lingnan. But circumstances had never allowed him to appreciate the landscape there; rather, adversity had always accompanied him. In 819, he finally had the time and the appropriate mood to fully enjoy what the famous scenic sites in Qujiang could offer. Therefore, contained in Han's simple, plainly worded request for a map-guide was a heartfelt joy and relief.

To conjecture a bit further, Han may even have hoped to find records about his brother in the map-guide, as it would record not only the geographical features but also the historical development of the place. It was conventional for map-guides to include information about the local governors' contributions to local affairs. In the *Map-Guide of the Shazhou Military Commandery*, for example, we can find records of four prefects of Shazhou: Li Wukui 李無亏 (?–694), Chen Xuangui 陳玄珪 (fl. 690s), Li Zulong 李祖隆 (fl. 680s), and Du Chuchen 杜楚臣 (fl.710s). Li Wukui's role as a capable, responsible governor is particularly highlighted in many entries about local landmarks.[84] The fact that Han's brother, Han Hui, as prefect of Shaozhou, may well have appeared in the local map-guide adds to Han's borrowings an implicit layer of personal affect. If this is the case, then Han's request for a map-guide also indicated his willingness to finally come to terms with his traumatic past through the mediation of a local geographic work.

Another contemporary poem titled "On the Tombstone of the Interior Official Cai in Suanzao County" ("Suanzao xian Cai Zhonglang bei" 題酸棗縣蔡中郎碑) by Wang Jian 王建 (767–830) further supplements our reading of Han's work:

> Green moss covered the characters, dust buried the stone turtle
> Wind and rain weathered his "utterly wonderful, lovely words"[85]
> Had I not read it from the map guide
> No one would have known this was Cai Yong's tombstone
>
> 蒼苔滿字土埋龜
> 風雨消磨絕妙詞
> 不是圖經中舊見

無人知是蔡邕碑[86]

This poem shows how the poet's previous reading of the map-guide of Suanzao County helped him identity a landmark. Cai Yong 蔡邕 (133–192), a famous man of letters in the late Eastern Han, died about six hundred years before Wang's time. As the poem shows, time had nearly rendered the tomb a desolate ruin, which made the identification of the tomb all the more valuable. Wang thus voices the anxiety of inadvertently losing track of something significant, which would cause both disorientation in the present and misrecognition of meaningful traces from a continuous history. For Han Yu, the map-guide was a public history of a place and an annotated reminder of his very personal childhood memories, while in this case, the map-guide enabled Wang Jian to reconstruct a literary past shared by all men of letters.

These two works that mention the map-guides explicitly as a guide for sightseeing indicate that during the mid-Tang period, the map-guides not only circulated among local officials, but also were made available to literati residing outside a given locality. With our limited information, we can have little sense of exactly how these map-guides circulated among outsiders. They might have borrowed them while visiting the place, or sent for them over long distances. In the preface to "Records of Chengdu" ("Chengdu ji" 成都記) by the Late-Tang literati Lu Qiu 盧求 (date unknown), he recalls: "At first, the map-guides of Sichuan were complete, and so many people from the central court requested and copied them. The person in charge was so bothered that he curtailed the volume greatly and secretly burnt them" (*Xianshi Xishu tujing shenbei, chaoye zhishi du jisheng xielu, zhu ziwu zhe, busheng qifan, sui jinxue er qianfen zhi* 先是西蜀圖經甚備，朝野之士多寄聲寫錄，主茲務者，不勝其煩，遂盡削而潛焚之).[87] From this passage, we know that by the Late Tang it was common for literati to borrow and copy local map-guides, and the multiple copies certainly facilitated their circulation beyond the local space. Accordingly, map-guides must have provided inspiration for even more authors across the empire as they spread further from their place of origin.

Conclusion

As this chapter has shown, map-guides played several distinct and significant roles in the culture and politics of the empire after the Rebellion through the mid-Tang. In many ways, they provided literary writers with necessary grist for the creative mill. As a means of organizing encyclopedic local knowledge, they preserved various cultural histories and projected the spiritual world onto the local geographical plane. They also were instrumental in facilitating local officials' writings and activities in key public spaces such as governmental offices and shrines. On a more personal level, they served as guidebooks for the literati's sightseeing and explorations. As a result of the literati's utilization of and close familiarity with these map-guides and the information in them, the literature of the time shows a creative use of geographic details, the reestablishment of historical continuities that might otherwise have been lost, and the artful connection of various geographical and imagined spaces in such a way as to re-delineate and reimagine cultural and sociopolitical borders. In other words, map-guides helped writers not only to orient themselves in the world, but also to navigate through rich and varied landscapes full of individual memories and experiences, communal histories, political and practical challenges, and many natural wonders. Above all, these geographical works served to inform and guide the literary inventions that were created in response to the new environments, both cultural and physical, that these writers encountered in the far reaches of the empire.

Notes

1. For a concise overview of mid-Tang literature's advancements in representing southern lands, see Chen Yuwen 陳毓文, "Zhongtang wenxue de difanghua jincheng" 中唐文學的地方化進程, *Fujian sheke lutan (sheke jiaoyu ban)* 福建社科論壇 (社科教育版), 2007: 9–10. A great deal of scholarship has been devoted to the relationship between some major mid-Tang writers and the local places where they served as officials. Given these writers' powerful influence upon the local culture from the mid-Tang to the modern day, this topic has also become an important component in studies of regional cultural histories. For instance, searching for "Liu Zongyuan and Yongzhou," we find 150 articles in the China Academic Journals Full-text Database (CNKI). In the case of Han Yu, his writing and activities in Chaozhou are an important topic in both literary history and in the study of the culture of Chaozhou (*Chaoxue* 潮學). For an interesting study that involves both fields, see Yang Ziyi 楊子怡, *Han Yu cichao yu Su Shi yuhui bijiao yanjiu* 韓愈刺潮與蘇軾寓惠比較研究 (Chengdu: Bashu shushe, 2008).
2. See Lei Wen 雷聞, *Jiaomiao zhiwai--Sui Tang guojia jisi yu zongjiao* 郊庙之外—隋唐国家祭祀与宗教 (Beijing: Sanlian shudian, 2009), 265–6.
3. In the early Tang, Xue Ji's 薛稷 (649–713) "Encomium Accompanying the Portraiture of Hermit Zhu" 朱隱士圖讚 quotes the *Map-Guide of Lingchi County* 靈池縣圖經 to tell the life story of Zhu Taochui 朱桃椎, a famous hermit and semi-mythical figure of the Sui Dynasty. See *Quan Tang wen*, 275.2799. In the High Tang, Zhang Yue's 張說 (667–731) "Biography of Duke Guo, Minister of the Military Department and Duke of Dai, Who Was Bestowed the Title of Junior Guardian" 兵部尚書代國公贈少保郭公行狀 tells how Guo Yuanzhen 郭元振 (656–713) consulted a map-guide to assist his military campaign in the Pamir Mountains. Ibid., 233.2355. In addition, the writer Li Shi's 李湜 (fl. 741) "Stele Inscription on the Chongyang Abbey of the Tang in Jiangzhou" 唐江州沖陽觀碑 mentions that the local landscape surrounding the Chongyang Abbey was so beautiful that "it was recorded in the map-guide several times" 圖經之載數矣. Ibid., 364.3702.
4. For some recent studies of office inscriptions, see Wu Heqing 吳河清, "Lun Tangdai tingbi jí de wenxian jiazhi" 論唐代廳壁記的文獻

價值, *Henan daxue xuebao (shehui kexue ban)* 河南大學學報 (社會科學版), vo1.51, no.3 (May 2011): 108–113. Chen Ke 陳可, "Tangdai tingbi ji yanjiu" 唐代廳壁記研究 (M.A. thesis, Shoudu shifan daxue, 2011). Liu Xingchao 劉興超, "Lun Tangdai tingbi ji" 論唐代廳壁記, *Sichuan daxue xuebao (zhexue shehui kexue ban)* 四川大學學報 (哲學社會科學版), no.3 (2008): 133–137. The historian Alexei Ditter distinguishes orisons of appeal from sacrificial orations (*jiwen* 祭文): "Sacrificial orations typically addressed the deceased friends or family members of their author, whereas orisons and vows nominally addressed divine beings." See Ditter, "Genre and the Transformation of Writing in Tang Dynasty China (618–907)," Ph.D. dissertation, Princeton University, 2009, 153. Some modern scholars have used the term *jinshen wen* to refer to this genre and thereby differentiated it from *jiwen*. For an overview of *jinshen wen*, see Yan Chunhua 嚴春華 "Tangdai jishen wen lunlüe" 唐代祭神文論略, *Hengyang shifan xueyuan xuebao* 衡陽師範學院學報, vol. 33, no. 2, 2012.4: 79–83

5. That the terms map-guides and map-records referred to the same geographical genre has been generally agreed upon among scholars of historical geography. See Lai Xinxia 來新夏, *Zhongguo difangzhi* 中國地方志 (Taipei: Shangwu yinshuguan), and Song Xi 宋晞 *Fangzhi xue yanjiu luncong* 方志學研究論叢 (Taipei: Taiwan shangwu yinshuguan, 1999), 172.
6. The abbreviation of *tujing* as *tu* can also be found in Song allusions to Tang map-guides. As Hua Linfu demonstrates, *Imperial Geography of the Taiping Xingguo Era* alludes to both *The Map-Guide of Jizhou* (*Jizhou tujing* 冀州圖經) *and The Map of Jizhou (Jizhou tu* 冀州圖), and in fact they are the same book. See Hua, "*Sui Tang tujing jikao shang*," 200–201.
7. For Yan's life, see Zhu Guantian 朱關田, *Yan Zhenqing nianpu* 顏真卿年譜 (Hangzhou: Xiling yinshe chubanshe, 2008), see also Amy McNair, *The Upright Brush: Yan Zhenqing's Calligraphy and Song Literati Politics* (Honolulu: University of Hawai'i Press, 1998), which is a comprehensive study of the reception history of Yan as a calligrapher, Confucian martyr, and cultural figure in various forms.
8. During the Song Dynasty, his inscriptions, largely in the form of rubbings, were carefully studied by some influential cultural elites such as Ouyang Xiu (1007–1072) 歐陽脩 and Su Shi (1037–1101) 蘇軾, who were instrumental in consolidating Yan's calligraphic style as the appropriate model for literati for many centuries to come. See McNair, *Upright Brush*,

especially Chapter Four: Partisan Politics at the Postrebellion Court, 60–82.

9. For example, Yan's record for Maid Ma's altar has been studied by historians of science as reflecting ancient Chinese people's awareness of geological changes or the use of fossils as evidence. See Needham, *Science and Civilization in China*, vol. 3, 599; see also Liu Shaomin 劉昭民 and Ruan Weizhou 阮維周, *Zhonghua dizhi xueshi* 中國地質學史 (Taipei: Shangwu yinshuguan, 1985), 36. Yan's writing about Lady Wei and her followers have been discussed by scholars such as Edward Schafer, Isabel Robinet, Suzanne Cahill, Yoshikawa Tadao, James Robson, Russell Kirkland, and N. Harry Rothschild.
10. As McNair points out, Yan was not a practicing Daoist himself, and so his efforts in favor of Daoism as both an official and a writer did not represent his own religious beliefs. McNair, *The Upright Brush*, 93–95.
11. Translation by Robert Campany with modifications, see *To Live as Long as Heaven and Earth: Ge Hong's Traditions of Divine Transcendents*, 262. As Joseph Needham has pointed out, "mulberry fields" 桑田 might have originally been a place name or the name of a constellation, but by the Tang it had the established connotation of "land which had once been covered by the sea, or would be in the future." Needham, *Science and Civilization in China*, vol. 3, 599.
12. *Shenxian zhuan*, ed. Ge Hong 葛洪 (284–364), in *Zengding Han Wei congshu* 增訂漢魏叢書 ed. Wang Mo 王謨 (ca. 1731–1817) (wood-block edition of 1791), 7.4a.
13. English translation is based on McNair's version in *The Upright Brush*, 85–86, with minor modification.
14. *Quan Tang wen*, 338.3424.
15. Hua, "Sui Tang tujing jikao xia," 82.
16. Ibid., 27.
17. Ibid., 19.
18. See Liao Yifang, *Tangdai de lishi jiyi*, 530.
19. Ibid., 523–530. In this respect, a comparable case is provided by Yan's friend Yuan Jie in "Record of the 'Grand Banquet Pavilion'" 廣宴亭記: "So I observed the topography and compared it with the map-guide. This place was verified as the site of the banquets of Wu. The county magistrate Mr. Ma climbed it and proclaimed, "Lord Xie's 'Poem for Fu Wuchang' says 'On Fan Mountain we held a grand banquet'." Doesn't that refer to this place? So I would like to build a pavilion according to this line, and call it "Grand Banquet Pavilion." What do you

think?" 乃相其地形，驗之圖記，津實為宴遊之處。縣大夫馬公登之，歎曰："謝公《贈伏武昌詩》云'樊山開廣宴，'非此地耶？吾欲因而脩之，命曰'廣宴亭，'何如？" See Yuan, *Yuan Cishan ji* 元次山集 (Shanghai: Zhonghua shuju, 1927), 9.5a. A fragment of *The Map-Guide of Jiangxia* (*Jiangxia tujing* 江夏圖經) includes exactly the same line by Xie Tiao 谢朓 (464–499): "The Fan Mountain, in the west. On land it is one hundred seventy-three *li* to the prefecture. The land yields amethyst. Several dozen paces to the east of the mountain there is a ridge. The space on the ridge is flat and wide, where green pine trees and bamboos always flourish. Under the ridge there is a creek that emits a chilly breeze, and therefore it is called the Chilly Creek. There is also a rock with dragon-shaped veins. Xie Xuanhui's poem says, 'On Fan Mountain we held a grand banquet.' That's the place." 樊山，西。陸路去州一百七十三里，出紫石英。山東數十步有岡，岡上甚平敞，青松綠竹，常自蔚然。其下有水溪，凜凜然常有寒氣，故謂之寒溪。有蟠龍石，謝玄暉詩云，"樊山開廣宴"，是也。Hua, "Sui Tang tujing jikao xia," 18.

20. Li, *Guben Dunhuang xiangtuzhi bazhong jianzheng*, 52.
21. McNair, *The Upright Brush*, 86.
22. *Quan Tang wen*, 338.3424.
23. In this ritual, "a golden image of a dragon was ritually thrown into a waterway to take messages to the gods." See Livia Kohn ed., *Daoism Handbook* (Leiden: Brill, 2000), 343.
24. McNair, *The Upright Brush*, 86.
25. *Quan Tang wen*, 338. 3424.
26. *Quan Tang wen*, 340.3453.
27. Ibid., 340.3454.
28. This genre originated in the early Tang and was widely popularized after the Rebellion. Among the forty-eight Tang authors whose writings in this genre are extant, forty were active during the mid-Tang including a stellar list of famous writers such as Gu Kuang 顧況 (c. 725–814), Liang Su 梁肅 (753–793), Quan Deyu, Han Yu, Liu Zongyuan, Liu Yuxi, Bai Juyi, Yuan Zhen, Lü Wen, Li Deyu, Huangfu Shi 皇甫湜 (777–835), and Shen Yazhi 沈亞之 (781–832). See Chen Ke, "Tangdai tingbi ji yanjiu," 24.
29. Xin, "Tangdai de dilixue," 443.
30. *Quan Tang wen*, 481.4917.
31. Liu was at the time serving as the Commandant of Lantian County, a nearby administrative district some 200 *li* from Wugong County. He was invited to draft the inscription because his friend in Wugong County

trusted his literary talent, rather than his geographic knowledge about the region.

32. The translation of this sentence is based on H.T. Huang's translation. Huang is a scholar of the history of the soybean in East Asia. See William Shurtleff, H.T. Huang, Akiko Aoyagi, *History of Soybeans and Soyfoods in China and Taiwan, and in Chinese Cookbooks, Restaurants, and Chinese Work with Soyfoods outside China: Extensively Annotated Bibliography and Sourcebook including Manchuria, Hong Kong and Tibet (1024 BCE to 2014)* (Soyinfo Center, 2014), 99.
33. Here Liu echoes the poem "Birth of the People" 生民 (from the "Greater Odes" 大雅 of the *Book of Songs*) 詩經, which held canonical authority for Confucian elites. Here is Arthur Waley's translation: "Indeed, the lucky grains were sent down to us, / The black millet, the double-kernelled, / Millet pink-sprouted and white. / Far and wide the black millet, the double-kernelled / He reaped and acred; / Far and wide the millet pink-sprouted and white / He carried in his arms, he bore on his back. / Brought them home, and created the sacrifice." See *The Book of Songs: The Ancient Chinese Classic of Poetry*, 246. 誕降嘉種，維秬維秠。維穈維芑，恒之秬秠。是穫是畝，恒之穈芑。是任是負，以歸肇祀。See *Shijing zhuxi*, 805.
34. *Liu Zongyuan ji jiaozhu*, 26.1724.
35. Ibid., 1724.
36. He, *Yimen dushu ji* 義門讀書記 (Beijing: Zhonghua shuju, 1987), 642. For more about the social stability of the capital area during the Tang, see Yang Yuejun 楊月君, *Tangdai jingji diqu zhian guanli yanjiu* 唐代京畿地區治安管理研究 (Beijing: Zhongguo shehui kexue chubanshe, 2009).
37. See Liao Yifang, *Tangdai de lishi jiyi*, 454. Chen Yunfeng 陳允鋒, "Lun Liu Yuxui de dili yishi jiqi shixue sixiang tedian" 論劉禹錫的地理意識及其詩學思想特點, in *Ningxia shehui kexue* 寧夏社會科學 (2004.5), 106–110.
38. Liu, *Liu Yuxi quanji biannian jiaozhu*, 6.345.
39. Zhang Ji张籍, *Zhang Ji ji xinian jiaozhu* 張籍集繫年校注 (Beijing: Zhonghua shuju, 2011), 4.539.
40. See Wang Guoying 王國瓔, *Zhongguo wenxue shi xinjiang* 中國文學史新講 (Taipei: Lianjing chuban gongsi, 2006), 52.
41. Jordan Alexander Gwyther, "Bai Juyi and the New Yuefu Movement," Master Thesis, University of Oregon, 2013, 150.
42. Bai, *Bai Juyi shiji jiaozhu* 白居易詩集校注, edited and annotated by Xie Siwei 謝思煒 (Beijing: Zhonghua shuju, 2006), 4.443.

43. Liu, *Liu Yuxi quanji biannian jiaozhu*, 6.345.
44. The earliest extant record of the story of Liyang from the *Huainanzi* of the Han Dynasty is as follows: "The city of Liyang was turned into a lake. The brave, the powerful, the sage, the intelligent, the weak, the timid, and the wretched all shared the same fate" 夫歷陽之都，一夕反而為湖、勇力聖知與罷怯不肖者同命。He Ning 何寧, *Huainanzi jishi* 淮南子集釋 (Beijing: Zhonghua shuju, 1998), 2.160. One example of a contemporary literary work that engages with this legend is Dugu Ji's 獨孤及 "The Rhapsody of a Dream of Traveling Afar" 夢遠遊賦: "I recognized my old country's cities and towers. With thousands and thousands of gates and doors, they looked like ant nests from afar. Hundreds of rivers flowed like silk. The five sacred mountains spread out. I searched for the old mountains and tall trees, but they were scattered and flickered in and out of sight. I saw the open roads of Yichuan (in modern Henan) had been trampled by barbarians; half of my old friends in Liyang have become fish and turtles" 識故國之城闕，千門萬戶，遙如蟻穴，百川綺分，五嶽羅列。覓舊山與喬木，纔依希而明滅。見伊川大道鞠為戎狄，曆陽故人半作魚鱉. The work tells of a Rebellion refugee's imaginary journey back home and the passage quoted depicts the view of his former home, now trampled by the barbarians. Dugu, *Piling ji jiaozhu* 毘陵集校注, Liu Peng 劉鵬 and Li Tao 李桃 eds. (Anshan: Liaohai chubanshe, 2006), 1.2. A more famous example that involves the legend of Liyang is found in Li He's poem "Do Not Go Out of Your Gates, Sir" 公無出門: "Though I have a horse to ride, I cannot go home. For the waves that drowned Liyang loom large as mountains, / Poisonous, horned dragons glaring, / rattling their brazen rings. / Lions and griffons drooling from slavering jaws" 我雖跨馬不得還，曆陽湖波大如山。毒虯相視振金環，狻猊猰貐吐饞涎。Li, *Li Changji shige biannian jianzhu*, 5.628. I use Frodsham's translation of Li He's poems with minor modifications. *The Poems of Li Ho*, 211.
45. *Liu Yuxi quanji biannian jiaozhu*, 6.346.
46. Ibid., 6.347.
47. Ibid., 5.340.
48. Ibid., 6.351.
49. See Hua Linfu, "Sui Tang tujing jikao xia," 63.
50. The legend can be found in Ren Fang's 任昉 (460–508) *Notes on the Extraordinary* (*Shuyi ji* 述異記): "Hezhou, Liyang County. Liyang Lake is thirty li to the west of the county. There was once a scholar

who met an old woman, and she treated him well. The scholar told her, 'The stone turtle at the county gate bloodied my eyes, so this land will be submerged into a lake.' The old woman went to see it several times, and the gatekeeper asked her why, and she told him. The gatekeeper dotted vermilion around the turtle's eyes. The old woman saw it and went to the northern mountains. Then the county was submerged" 昔有書生遇一老姥，姥待之厚，生謂姥曰: 此縣門石龜眼血出，此地當陷爲湖。姥數往視之。門吏問姥，姥具答之。吏以朱點龜眼。姥見之，遂走上北山，顧城遂陷焉。Ren, *Shu yiji* (Beijing: Zhonghua shuju, 1960), 1.10.

51. Liu, *Liu Yuxi quanji biannian jiaozhu*, 6.353.
52. Ibid., 6.345, 16.1087.
53. Ibid., 16.1088.
54. Many scholars have discussed local folk cults during the Tang-Song period in China, including Sue Takashi, Gao Mingshi, Gan Huaizhen, Howard Wechsler, Terry Kleeman, Jean Levi, Valerie Hansen, and Lei Wen. Since Lei Wen's scholarship has synthesized much of the previous scholarship, and discussed the function of map-guides in the Tang government's management of local shrines in particular, I will mainly draw on his work *Jiaomiao zhiwai*.
55. Lei, *Jiaomiao zhiwai*, 255–258.
56. Ibid., 263.
57. Ibid., 262.
58. Ibid., 262–270.
59. Li, *Guben Dunhuang xiangtuzhi bazhong jianzheng*, 52.
60. Hua, "Sui Tang tujing jikao xia," 5–6.
61. Yuan, *Yuan Cishan ji*, 10.7a.
62. Liu, *Liu Zongyuan ji jiaozhu*, 28.1836.
63. *Liu Zongyuan ji jiaozhu*, 28.1836.
64. In Sima Qian's biography of Qu Yuan, one of Qu's major enemies is referred to as Grand Master Shangguan (*Shangguan dafu* 上官大夫). See *Shiji*, 84.2481. Many traditional and modern commentators on Qu's poetry have identified him with Jin Shang, an official trusted by King Huai of the Chu Kingdom. Scholarship in recent decades has demonstrated that Grand Master Shangguan and Jin Shang were two different persons. See Fan Zhengsheng 范正聲, "Jin Shang bushi Shangguan dafu" 靳尚不是上官大夫, *Tai'an shizhuan xuebao* 泰安師專學報, 1994.4: 407–408, and Li Weishi 李偉實, "Shangguan dafu he Jin Shang shi

liangge ren" 上官大夫和靳尚是兩個人, *Linglin shifan gaodeng zhuanke xuexiao xuebao* 零陵師範高等專科學校學報, 1999.2: 33–34

65. *Quan Tang shi*, 474.5376.
66. Hua, "Sui Tang tujing jikao xia," 24.
67. Lei, *Jiaomiao zhi wai*, 246–250.
68. Li, *Li Deyu wenji jianjiao, waiji* 4.694.
69. Liu, *Liu Zongyuan ji jiaozhu*, 41.2621.
70. Yuan, *Yuan Zhen ji biannian jianzhu sanwen juan*, 277.
71. Bai, *Bai Juyi wenji jiaozhu* 白居易文集校注 ed. Xie Siwei 謝思煒 (Beijing: Zhonghua shuju, 20110), 3.155–56.
72. David McMullen observes that "the texts the code supplies are notable for the laconic quality: they usually involved the participant stating the year, month and day, identifying himself, and then giving a very brief address in archaic and honorific language." See McMullen, "Bureaucrats and Cosmology: The Ritual Code of T'ang China," in David Cannadine and Simon Price, eds. *Rituals of Royalty: Power and Ceremonial in Traditional Societies* (Cambridge: Cambridge University Press, 1987), 201–202.
73. For a discussion of Liu's piece, see Ditter, "Genre and the Transformation of Writing in Tang Dynasty China (618–907)," 197.
74. Ibid., 126.
75. Ibid., 127.
76. "Genre and the Transformation of Writing in Tang Dynasty China (618–907)," 197.
77. Now Shaoguan in Guangdong Province, during the Tang Dynasty, Shaozhou belonged to the Lingnan Administrative District (Lingnan Dao) and its government office was founded in the region of Qujiang.
78. Fang Shiju 方世舉, *Han Changli shiji biannian jianzhu* 韓昌黎詩集編年箋注 (Beijing: Zhonghua shuju, 2012), 11.587.
79. See Liu Guoying 劉國盈, *Han Yu pingzhuan* 韓愈評傳 (Beijing: Beijing shifan xueyuan chubanshe, 1991), 13–15.
80. Han, *Han Changli wenji jianzhu* 韓昌黎文集校注, collated and annotated by Ma Qichang 馬其昶, (Taipei: Shijie shuju, 1967), 1.3
81. Lü Dafang 呂大防, *Han Yu nianpu* 韩愈年譜 (Beijing: Zhonghua shuju, 1991), 68.
82. Han, *Han Changli shiji biannian jianzhu*, 10.584.
83. The following discussion of Han's trip is based on Ke Wancheng's 柯萬成 detailed study of Han's poems composed on the trip. See Ke, "Han Yu bianchao xingji yu sanshi jinian xinlun" 韓愈貶潮行跡與三詩繫年新論, in *Han Yu yu Lingnan wenhua* 韓愈與嶺南文化, eds. by Zhang

Qinghua 張清華, Hu Axiang 胡阿祥, and Liu Yingjie 劉英傑 (Beijing: Xueyuan chubanshe, 2002), 370–386.

84. For discussions of how these governors feature in the *Map-Guide of the Shazhou Military Commandery*, see Zhao Zhen 趙貞, "Lun Tangdai tujing de bianxiu" 論唐代圖經的編修, *Shixuan shi yanjiu* 史學史研究, 2013.4, 95–96.
85. The "utterly wonderful, lovely words" is an allusion to Cai Yong's appraisal of the Han Dynasty "A Stele Inscription for Maiden Cao" 曹娥碑, written on the back of the stele. It was said the appraisal was given in the form of a riddle: "yellow pongee, young wife, maternal grandson, ground in a mortar" 黃絹幼婦，外孫齏臼. The solution is "utterly wonderful, lovely words" 絕妙好辭. See Wang, *Wang Jian shiji jiaozhu* 王建詩集校注, edited and annotated by Yin Zhanhua 尹占華 (Chengdu: Bashu shushe, 2006), 9.407.
86. Ibid., 407.
87. *Quan Tang wen*, 744.7703.

Chapter 4

Into the Deep South

The Aesthetics of New Landmark Creation

As the previous two chapters have shown, grand maps and map-guides helped shape the literati's understanding of the empire and of local spheres, enriching their literary imagination of various cultural and geopolitical spaces. This chapter explores how literature went beyond the purview of geographic works to represent the empire's uncharted landscape in the deep south. It examines how Yuan Jie, Liu Zongyuan, Han Yu, and Liu Yuxi captured a radical geographic transformation in the empire propelled by the epic southward migration of the Chinese population. These writers created new literary amalgams, combining representations of geographical features, lyrical evocation, and unique visions of the expanding medieval world. Conversely, their literary production was also integral to a collective imperial project of discoveing, demarcating, and developing local spaces. Their naming and inscribing of local landmarks profoundly transformed the natural landscape, offering new material to later geographic works. This cycle of mutual enrichment between literature and geography will be the focus of my discussion here.

The central texts in question are landscape essays. In this chapter I follow Xiaoshan Yang in defining the term landscape essay as "prose writings wherein descriptions of natural scenery are structurally and thematically indispensable." This definition includes not only accounts of excursions (*youji* 遊記), but also inscriptions (*ming* 銘) and prefaces (*xu* 序) to the inscriptions.[1] The genre name *youji*, commonly translated as travel records or accounts of excursions, is a term modern scholars have coined and applied retrospectively to the Tang.[2] Although literary representations of travel and landscapes appeared much earlier in Chinese history, modern literary historians generally attribute the formation of the genre conventions to Liu Zongyuan. His famous "Eight Records of Yongzhou" ("Yongzhou baji" 永州八記) is arguably the founding example of accounts of excursions.[3] Limited in number, this small body of works nonetheless contributed enormously to the development of classical Chinese prose, leaving a lasting influence on literary and cultural history.[4] In addition to accounts of excursions, Yang's definition also includes inscriptions and prefaces to the inscriptions. These are records written for specific scenic sites, often with the expectation that the text would be engraved either in a stele or on a rock at the site. Once inscribed, they became an organic part of the landscape for centuries to come. This broad definition of landscape essays works nicely for my discussion because, as I will show in the following pages, the various genres of texts conforming to this definition, in their various material forms, provide us with a full picture of the interplay between literature and geography in the deep south at this complicated juncture of imperial history.

These landscape essays—Liu Zongyuan's masterpieces in particular—have been a favorite subject of analysis for both premodern critics and modern scholars. Much has been written about their aesthetic value, cultural meaning, and political implications. Departing from this interpretative tradition, I recontextualize the essays within the history of migration following the Rebellion, a transformative process that was vital in shaping the authors' understanding of the empire. This shift in context allows us to appreciate a common theme seen in many landscape essays of the time, as these landscape essayists, whether celebrated or lesser-known, all strove to create the central spatial imaginary of their new home in the deep south. This new home may have been a temporary shelter, a contingent place of abode, or a permanent settlement; regardless, the spatial imaginary always suited the taste of the relocated northern elites. My interdisciplinary approach to the landscape essay thus adds significantly to previous studies by providing a geographically contextualized reading of the literary texts and aims to provide a deepened understanding of the relationship between text and space.

I use the term "deep south" to denote the geographical focus of the chapter, namely the southern regions of the Tang Empire that remained economically underdeveloped before the mid-eighth century. The Lingnan region, which is roughly coterminous with present-day Guangdong, Guangxi, and Hainan provinces, along with northern Vietnam, constitutes the largest chunk of the area.[5] The reach of the deep south also extends to today's Hunan provinces. I use the modifier "deep" to differentiate the area from the affluent Jiangnan region (literally "south of the river"), the populous lower Yangtze areas that buttressed the Tang with their continued flow of tax revenue after the Rebellion. The deep south, in contrast, remained significantly underdeveloped when compared to the Jiangnan region. The term "deep" also serves to highlight its remoteness from the imperial center and relative unfamiliarity to men of letters. As the Rebellion threw the central plains into turmoil, however, the region saw exponential migrant population growth and impressive development.

As for the writers who wrote in the deep south, I highlight their identity as uprooted officials in search of a new home, as this identity is featured prominently in their works. Most of them were transferred to undesirable posts in the deep south after political misfortune in the capital. They saw their banishment as temporary, and they waited anxiously for a chance to return. This did not stop them from seeking some sort of home in the deep south, sanctuaries that could afford a much-needed sense of safety and sensory pleasure. On another level, these landscape essayists, serving as local officials, were acutely aware that the southern land was no longer an unchanging wilderness, but rather was a shifting ground undergoing rapid transformation. This dynamism offered writers plenty of opportunities to acquire new lands, build new houses, and take pleasure in their explorations of a fresh landscape. It also sometimes heightened their responsibility as imperial agents to facilitate communal dwelling in regions under their supervision. In any case, the issue of dwelling place was central to their lives in the deep south and was handled with both enthusiasm and subtlety in their landscape writing.

To be able to speak of dwelling in the context of relocation and migration, we need to define it in a nuanced way. For these landscape essayists, the sense of an ideal home did not necessarily imply a permanent stay. In this regard, Western literary scholars have offered useful insights in recent decades. For instance, David Seamon, in discussing the motif of dwelling in novels representing Swedish emigrants in America by Vilhelm Moberg, considers dwelling as the dialectical opposition to journey. According to Seamon, a journey carries a person away from a stable home, whereas dwelling relates to the basic human need for spatial order and familiarity. Following Heidegger, who considers dwelling the way "we human beings *are* on the earth," Seamon characterizes dwelling in a migration context as a status of deep rest that "involves a lifestyle of regularity, repetition and cyclicity all grounded in an atmosphere of care and concern for places, things, and people."[6] He also points out that since a pure form of dwelling is unlikely to be possible in practice, what really matters are human efforts to learn to dwell and to seek better ways in which to dwell.[7] These efforts were at the core of mid-Tang writers' explorations of the deep south. As is reflected in their writings about local landscapes, the natural scenery they appreciated and the new landmarks they built almost always embodied a quest for a more comfortable place to dwell and to feel at home.

On the other hand, it is necessary to differentiate the elite migration to the deep south and the broader southward demographic flow during and after the Rebellion. As Nicolas Tackett points out in his book *The Destruction of the Medival Chinese Aristocracy*, the post-Rebellion elite migration "should not be thought of as part of the great southward demographic shift that spanned the eighth through thirteenth centuries; the Tang capitals remained a magnet for elites until the fall of the dynasty."[8] According to Tackett, most elites relocating to the south sought any opportunity to return to the capital region. Some managed to do so while alive; others who died in the south at least had their tombs relocated to the north by their families. Therefore, I am not arguing that these elites were simply members of the southward wave of migration, distinguished only by the fact that they wrote about their travels; rather, they had different trajectories from other migrants, even as they witnessed, interacted with, and were inspired by the lives of the latter. The elite writers' distinct trajactories were reflected in their writings. First, because of the frequent loop of relocation and return among elite families, many essayists to be discussed in this chapter first experienced the south with their families during childhood, providing a background that was often evoked in their later southward official travels. Second, the lack of a sense of permanent relocation paradoxically fueled their passion to seek inhabitable spots in many places; finding home nowhere and finding it anywhere are two sides of the same coin.

My investigation of the mid-Tang landscape essay in relation to geography brings a fresh understanding to these well studied literary works. In the large body of critical assessments on the landscape essay, the prevalent approach has been to see the formation of the genre as a consequence of the authors' political banishment. Critics in this tradition tend to read these writers' explorations of the aesthetically satisfying natural environment as their way of seeking comfort for themselves or achieving a philosophical introspection into their afflicted inner selves.[9] Thus interpreted, the landscapes of the periphery also become a metaphor for the marginalized authors themselves.[10] The approach of taking these works as banishment literature, or *bianzhe wenxue* 貶謫文學, works nicely with the shared exilic experiences of the majority of authors to be discussed in this chapter, with the possible exception of Yuan Jie.[11] This approach, however, generally assumes that the southern landscape was a static, hidden corner of nature waiting to be discovered and appreciated by the solitary landscape essayist, and thus fails to take into account the geographic dynamism of the land. Consequently, this way of reading focuses primarily on the political and moral implications of the landscape essay on a personal level. My analysis, by reframing the landscape essay against a broad and dynamic geographic backdrop, explores the deeper meanings of the texts, as well as their significance in terms of the large-scale social changes of the time.

In addition, in my analysis of carefully selected works, I hope to capture the decisive moments of literary creativity that distinguish literature from geographic records. In chapters 1 and 3, we compared landscape essays with the geographic works from which they drew inspiration, such as land records and map-guides, and found that in fact the two types of writing share many stylistic similarities in terms of landscape description. Accordingly, the essential difference between literary and geographic representation of landscape lies in the different human vs. nature relationships that their spatial imaginaries entertain. Geographic writings such as land records and map-guides assume the tone of an intellectual or official surveyor who describes the landscape in question as an outsider; in these works, the unique natural beauty of the place is represented as a boon to the geographic features that happen to be especially suited to the literati's aesthetic tastes. In contrast, the landscape essayist tells the story of human exploration, marking out an evolving geography in which newly established landmarks significantly change the spatial order of the local environment. The hero of the latter is not only a surveyor of the land but also an active participant in landscape making. By highlighting this component of human participation, my study of the landscape essay crystalizes the creative abundance found in those works beyond their geographical references. It is precisely this abundance that defines their literary value.

In what follows, I will first give a brief overview of the massive transformation that the deep south underwent during the post-Rebellion migration. This turbulent scene of refuge and relocation was represented in creative ways in writings by landscape essayists. I will then take the pioneer landscape essayist Yuan Jie as a case study to illustrate how this genre, in its post-Rebellion inception, integrated Yuan's personal memory and experience as a migrant with his ambition for communal development as a local administrator. The chapter will then deal with the mid-Tang writers who carried on Yuan's legacy and further developed it into a collective project that entailed various forms of communication and collaboration within the community of literati who served local governments in the deep south. The end of the chapter focuses on Liu Zongyuan, whose works delicately manifested the author's geographical understanding of the empire and his estimable art of dwelling.

Migration and Relocation in the Deep South

In the history of demography, the time period during and after the Rebellion marks one of the most radical changes in premodern China, as the long-lasting social chaos redistributed a great northern population towards the south. The Rebellion alone contributed around 2.5 million southward-moving migrants.[12] In the years immediately following the Rebellion, a temporary stability in the north attracted a number of migrants, mostly those with higher social status, back to their hometowns. Regional wars in the 780s caused by rebellious military governors, however, yet again drove northerners south. The reduction of population in the north resulted in heavier taxation on those who stayed, thus initiating another round of migration for financial reasons. Although statistics for migration across the entire country in the mid-Tang are not available, many records in *Complete Tang Prose* and *The Old History of the Tang* testify to the historical change on a regional scale.[13] The process of migration brought the unfamiliar deep south more fully into the purview of the empire, as it was simultaneously considered an underdeveloped backwater and a beautiful and comfortable land in which to settle.

The enormous and chaotic social changes during and after the Rebellion affected the landscape essayists and their writing in several ways. First, the increasing need for governing personnel in the south had driven several generations of Tang elites to the south. As a result, many of the essayists had already encountered the region in their younger years, and thus had early memories of the land firmly in their minds when they were relocated to the deep south as officials themselves. Yuan Jie, a pioneer in landscape writing, was a wartime refugee. Some mid-Tang writers, such as Han Yu and Liu Zongyuan, first went to the deep south in the company of their family members and their later political relocation in effect marks their return to that region. In other words, these writers' encounter with the deep south in middle age does not represent an abrupt rupture in their otherwise smooth lives, but it was rather part of a lifelong engagement with the journey southward in the context of broader social change. Accordingly, when they reencountered the deep south in their adulthood as officials, the challenge that they faced was not merely the backward living conditions, but also the sense of instability, insecurity, fear, and loss that they had associated with the deep south since childhood. Furthermore, to echo my definition of dwelling in a context of constant movement, when they dwelled in the deep south, what mattered was not whether it would be a permanent stay, but whether they could recreate a sense of security and feel at home according to a northerner's standards, in an environment that was otherwise uninviting.

Yuan Jie was born in Henan and became a Presented Scholar (*jinshi* 進士) degree holder in 754, but before he had a chance to take up an appropriate official post, the Rebellion broke out. An adept community leader, he led more than two hundred households southward to take refuge in several places in the mid-Yangtze area. A few years later, he established himself as a highly skilled official in both military affairs and civil governance in a series of important posts, most notably under the Jingnan military commissioner, whose command was held in present-day Hubei Province. The creation of the post of Jingnan military commissioner was itself a result of migration. According to *The Old History of the Tang Dynasty*, "Since the Zhide Reign (756–758), the central plain was restless, and the people of Xiang and Deng as well as the elites of the two capitals all fled to the Yangtze River and the Xiang River areas. With this, the population of Jingnan increased tenfold. The court accordingly created the post of a military commissioner of Jingnan" (*Zi Zhide hou, zhongyuan duogu, Xiang Deng baixing, liangjing yiguan, jintou Jiang Xiang, gu Jingnan jingyi, shibei qichu, naizhi Jingnan jiedushi* 自至德後, 中原多故, 襄鄧百姓, 兩京衣冠, 盡投江湘, 故荊南井邑, 十倍其初, 乃置荊南節度使).[14] The region under the Jingnan military commissioner played an increasingly important role as a major revenue source and military base during the war. As a consequence, in 760, the Tang government raised its administrative center, Jingzhou prefecture (Jingzhou 荊州), to a superior prefecture (*fu* 府) and renamed it Jiangling 江陵, designating it the empire's southern capital Nandu 南都.[15] The success that Yuan Jie had in this regional administration prepared him for his later career, including his post as the prefect of Daozhou (in today's southern Hunan Province), where he composed some of his most famous landscape essays.[16]

Liu Zongyuan's southern experience is much more complicated. Liu's father Liu Zhen 柳鎮 (?–793), like many social elites of his time, escaped the Rebellion to the lower Yangtze area. After the Rebellion, he returned to Chang'an for a short while with his immediate family, leaving the majority of his clan members in the south. As Jo-shui Chen points out, this was a common situation at the time, as a poem by the famous writer Sikong Shu司空曙 (740–790?) testifies: "When the world fell into turmoil, we went south together. / As time settles things, you return north alone" (Shiluan tong nanqu, shiqing du beihuan 世亂同南去，時清獨北還).[17] In the mid-780s, Liu Zhen again went to the south and his teenaged son Liu Zongyuan accompanied him. Between the ages of ten and thirteen, Liu Zongyuan traveled around the mid-Yangtze area for extended periods with his father, including his first encounter with the deep south in the Hunan region.[18]

In 815, when Liu was on his way to Liuzhou, he passed by a tower in Changsha, Hunan, which he had visited decades ago. What this reunion aroused in him is illustrated in the poem "Thinking of the Past on the Southern Tower in Front of the Changsha Post House (Here I Parted with the Venerable De)" ("Changsha yiqian nanlou ganjiu [xi yu Degong bieyu ci]" 長沙驛前南樓感舊 [昔與德公別於此]):

> After parting with the sea crane
> Thirty autumns later, life and death separate us
> Now I come again with a few lines of tears
> Ascending the southern tower of the post house alone.

> 海鶴一為別
> 存亡三十秋
> 今來數行淚
> 獨上驛南樓[19]

This poem describes how the adult Liu Zongyuan was suddenly struck by the memory of parting with an elder at the post-house in Changsha some thirty years before. As it turns out, that parting marked their permanent separation. As the poem shows, in 815, when Liu once again embarked on an unpleasant journey to Liuzhou in the Lingnan area as a political punishment, his past experience of separation in the post-house reinforced his melancholy mood upon returning to the deep south. Liu eventually died in Liuzhou, thus completing a cycle of loss and death, movement and rest in the deep south.

Han Yu was also deeply involved in the mass southern relocation of population during and after the Rebellion. He lost his parents in infancy and was brought up in his elder brother Han Hui's family. Han Hui fled to the Jiangnan area during the rebellion, where he became a famous intellectual and made political connections. In 774, he was appointed Diarist of Activity and Repose (*qiju sheren* 起居舍人) and started his career in the court. In 777, Han Hui was demoted to Shaozhou in Lingnan. Han Yu, then ten years old, followed him as a dependent. Han Hui died in Shaozhou merely two years after taking up the post. Han Yu then returned to the north with his sister-in-law in 779. In the early 780s, when another war broke out between the central government and the military governors of the Hebei region, Han Yu and his extended family of a hundred people were forced to flee to Xuancheng, in today's Anhui, where Han then lived for five years.[20]

Like Liu Zongyuan, Han Yu associates his early southern experience with hardship and loneliness in his writing, as we have seen in chapter 3 in the discussion of his works such as "Rhapsody on Recovering My Purpose" and "Expressing My Thoughts Passing the Mouth of the River in Shixing." For both Han and Liu, their revisit to the deep south as adults brings about a helpless lament over past losses that happened in that place. If the will to find a dwelling place and engage in homebuilding sits on one end of the spectrum of their lifelong southern experience, then the fear of death and insecurity marked the other.

In addition to bringing a migratory lifestyle to these landscape essayists, the social changes accompanying mass migration gave rise to an evolving trend of land consolidation, which provided the important material conditions for these writers to find and build new homes in the deep south. While the early Tang emperors supported an idealist blueprint for an egalitarian state-owned land tenure system, the successive rebellions had made its implementation impossible.[21] As the state withdrew from regulating the land market, men of means found a convenient way to ensure the sustained financial well-being of their families in the accumulation of landed property. This phenomenon was more pronounced in the north, where destitute peasants willingly surrendered their farms to anyone who could offer funds. The uprooted elites carried this collective social desire for landed property with them to the deep south. Wherever they settled, they bought lands and passed them down through the generations.

The zeal for land acquisition was as much about profit as it was about personal dwelling places, as many new landlords built extensive houses and gardens near their farming manors for personal enjoyment.[22] Some famous contemporary villas owned by literati-officials included Pei Du's Wuqiao Villa (*Wuqiao zhuang* 午橋莊) and Li Deyu's Pingquan Villa (*Pingquan zhuang* 平泉莊).[23] Liu Zongyuan's cousin Liu Mou, too, lived in a house surrounded comfortably by mulberry trees in a manor boasting five hundred *mu* of land in Jiangling in the mid-Yangzte area.[24] As the later discussion will show, Yuan Jie and Liu Zongyuan also had an eye for both good farmland and ideal dwelling sites, as well as an enthusiasm for either acquiring the places for themselves or marketing them to other potential buyers. In short, impressive economic development in the deep south translated seamlessly into an elite quest for local land, either as property or for homebuilding.

Finally, the mass migration to the deep south and the consequent radical shift of the geopolitical structure of the empire provided contemporary writers new ways to imagine the southern space and to engage with it in literature. An early example can be found in Zhang Shuqing's 張叔卿 (fl. 760) poem "Exile to Guizhou" ("Liu Guizhou" 流桂州):

> Don't ask whether Cangwu is too far away,
> Since the roads in today's world are so hard.
> Wherever the barbaric dust cannot reach
> We can just call it Little Chang'an.
>
> 莫問蒼梧遠
> 而今世路難
> 胡塵不到處
> 即是小長安[25]

Guizhou was located in the Lingnan area, a place that was less chaotic during the Rebellion and therefore received a great number of northern migrants. In Zhang's poem, the author compares the remote Guizhou to the imperial capital Chang'an only because "the barbaric dust cannot reach" it. Zhang's observation as an exile reflects a common sentiment of road-weary northern migrants who went wherever they could take shelter from the devastating war. From this perspective, the deep south was no longer represented as a barbarian land; quite to the contrary, it became a safe haven from the barbarians.

Among mid-Tang writers, Liu Zongyuan demonstrated a particular awareness of the social phenomenon of migration and the devastating impact it had on local lives in his essay “Discourse of the Snake Catcher” (“Bushe zhe shuo” 捕蛇者說). The essay recounts a dramatic and deeply troubling conversation between Liu and a certain Mr. Jiang who lived in Yongzhou. Three generations of the Jiang family chose the potentially fatal job of catching an unusual type of snake for the court as an ingredient in medicine. As Jiang told Liu, although both his grandfather and his father had died from snake bites, he still risked his life to catch snakes because doing so exempted him from the high land tax. Upon hearing Jiang tell his sad story, Liu laments the brutal social reality: “Confucius said, ‘Harsh government is fiercer than a tiger.’ I once wondered about the truth of that, but judging from Jiang’s case, it is still true. Alas! Who could have thought that the poison of tax collection is worse than that of a snake?”[26] (*Yu wen er yubei. Kongzi yue: “Kezheng mengyu hu ye.” Wu chang yihu shi, jin yi Jiangshi guanzhi, youxin. Wuhu! Shuzhi fulian zhidu, you shenshi shezhe hu!* 余聞而愈悲。孔子曰：“苛政猛於虎也。”吾嘗疑乎是，今以蔣氏觀之，猶信。嗚呼！孰知賦斂之毒，有甚是蛇者乎！)[27]

Embedded in “Discourse of the Snake Catcher” is the history of a migrant family in Yongzhou. Although in his writings Liu repeatedly called Yongzhou a “barbaric land,” Jiang as a man of Yongzhou (*Yong zhi ren* 永之人) was not an “aboriginal.” This is clear from elsewhere in Liu’s writings, where he describes the aboriginal inhabitants of the region as impossible to communicate with due to the language barrier. Liu thus retells Jiang’s narrative:

> It has been sixty years now since my family came to live in this place, over three generations ago. The livelihood of our neighbors has grown more difficult by the day. They have exhausted everything that their land produces, used up all the earnings of

> their houses, cried out for help, and moved away. Staggering and falling with hunger and thirst, they have braved the wind and the rain, the cold and the heat, breathed poisonous vapors, and frequently died falling on top of one another. Of those who lived in my grandfather's time, fewer than one in ten are left in their houses. Of those who lived in my father's time, fewer than two or three in ten are left in each family. Of those who have been living with me for the last twelve years, not four or five in ten are left in each family. They either died or move to other places. But I alone have survived because I catch snakes.[28]
>
> 自吾氏三世居是鄉，積于今六十歲矣。而鄉鄰之生日蹙。殫其地之出，竭其廬之入。號呼而轉徙，飢渴而頓踣。觸風雨，犯寒暑，呼嘘毒癘，往往而死者相藉也。曩與吾祖居者，今其室十無一焉。與吾父居者，今其室十無二三焉。與吾居十二年者，今其室十無四五焉，非死則徙爾。而吾以捕蛇獨存。[29]

Jiang's grandfather had moved to Yongzhou roughly sixty years earlier, that is, when the Rebellion broke out. The emotional intensity of Liu Zongyuan's lament is built upon the irony that the Jiang family migrated to Yongzhou to survive, but in order to survive they were forced to take up a deadly profession. As Jiang says, his neighbors "either died or moved to other places." Such tragedies further reveal the harsh social reality of the post-Rebellion age, which resulted in more and more death and displacement. Liu's sympathy towards the Jiangs' struggles for survival arises from his sense of social justice as a marginalized official, but also from his own experiences and personal understanding of dislocation.

To summarize, the broad social transformation that the deep south underwent right before and during the mid-Tang informed contemporary writers' landscape making and landscape writing in many ways. The social chaos led to their youthful exposure to the deep south, thus making the land a symbol of uncertainty and insecurity for them before they were demoted to the region as adults. Economically speaking, radical land consolidation facilitated their purchasing of land and homebuilding. This serves as an important background that will allow us to fully comprehend the landscape essays of Yuan Jie and Liu Zongyuan. Finally, the southern dynamism these writers found is explicitly addressed in their literary works, indicating that they were well aware of the social and cultural environment around them. In the following discussion, we will look at individual works by Yuan Jie, Han Yu, Liu Yuxi and Liu Zongyuan within the context outlined earlier.

Yuan Jie and the Transformation of Cangwu

Yuan Jie is considered by many critics to be a precursor of the mid-Tang landscape essayists. The Qing scholar Wu Rulun 吳汝綸 believed that "Cishan's [Yuan Jie] wild enjoyment of landscapes anticipated Zihou's [Liu Zongyuan] writing" (*Cishan fangzi shanshui, shikai Zihou xiansheng* 次山放恣山水，實開子厚先聲).[30] At the same time, Yuan was a former war refugee and a long-term local governor in the deep south. At the core of his landscape essays was an image of a dwelling place suitable for relocated northerners, especially northern elites like himself. On the one hand, Yuan was obsessed with naming and inscribing new scenic spots, attributing to them his own memories and values in the large wave of migration and settlement. On the other hand, his personal homebuilding was connected with a grand vision of imperial revival, most notably in his inscription of "Ode to the Revival of the Great Tang" ("Da Tang zhongxing song" 大唐中興頌), written in 771.

The region where Yuan produced most of his landscape essays was Daozhou. To investigate his interactions with local geography, I would like to first give an overview of the transformation of the region's social and literary image after the Rebellion. Daozhou, as an administrative district, did not garner much attention from the literati before the Rebellion. When writers referred to the larger region where Daozhou was located, they were more likely to call it Cangwu 蒼梧. Cangwu as a geographical concept was often associated with Jiuyi Mountain and the point of convergence where the Xiang River joined the Xiao River.

The Cangwu region stretched across most territories of Daozhou and Yongzhou. It was made famous by the myth of the ancient sage king Shun, who embarked on a long journey to the south and died in the wilderness of this region. Upon hearing the tragic news, his two loyal consorts Ehuang and Nüying committed suicide in the Xiang River and came to be worshipped as the river's patron goddesses. This ancient legend has never developed into a coherent story, leaving much empty space for conjecture and imagination.[31]

In Tang poetry, the name of Cangwu almost always carries the mythical associations of a distant place from which travelers may not be able to return home. As Li Bai's "Wailing for My Dear Chao Heng" ("Ku Chao qing Heng" 哭晁卿衡) has it:

> Chao Heng of Japan bid farewell to the imperial capital.
> The journeying sail circled Penghu Island.
> The bright moon will not return, sinking into the emerald sea.
> The white clouds' melancholy color covers Cangwu.

> 日本晁卿辭帝都
> 征帆一片繞蓬壺
> 明月不歸沈碧海
> 白雲愁色滿蒼梧[32]

Li wrote this poem to pay homage to his Japanese friend Chao Heng 晁衡 (a.k.a. Abe no Nakamaro 阿倍仲麻呂) after he heard that Chao's death had occurred on his way back to Japan, although the sad news later turned out to be inaccurate.[33] In the poem, the author associates Chao Heng with two geographic names, Penghu and Cangwu. Penghu refers to the mystical Penglai Island supposedly in the East Sea that we saw in Li's poem about mythical geography in chapter 2. During the Tang, Penglai Island became a metonym for Japan in both Chinese poetry and in the letters of Japanese envoys to the Tang court.[34] The place name Penghu therefore indicates Chao Heng's literal destination; simultaneously, it gives the poem an otherworldly atmosphere that merges with the evocation of death in the third line, "sinking into the emerald sea." Cangwu, on the other hand, is in Hunan and is not directly related to Chao Heng. Li evokes Cangwu to rhyme with Penghu and to echo the long tradition of mourning the dead traveler who is unable to return home.

Meng Haoran's "Seeing off Yuan the Tenth to Seek His Younger Brother in Lingnan" ("Song Yuanshi Lingnan xundi" 送袁十嶺南尋弟) employs the imagery of Cangwu in a more straightforward way. As the title shows, the poem was written for Meng's friend Yuan Shi 袁十 before he set out for Lingnan to visit his brother. Envisioning the rough journey that Yuan was about to embark upon, Meng wrote, "Far away are the white clouds of Cangwu. / Deep is the misty water of Dongting Lake" (*Cangwu baiyun yuan, yanshui Dongting shen* 蒼梧白雲遠，煙水洞庭深).[35] In this couplet, Cangwu is paired with Lake Dongting as two synecdoches for the remote south. In both Li and Meng's poems, the white clouds rather than actual features of the land become the symbol of Cangwu, further highlighting the mythical atmosphere associated with the place.

Immediately after the Rebellion, the blurry, mystic realm of Cangwu was transformed into the defined geographic region of Daozhou, largely thanks to Yuan Jie's persistent engagement with it in his writing. In 763, Yuan was appointed prefect of Daozhou. Due to the influx of migrants, the population of Daozhou doubled after the An Lushan Rebellion.[36] Soon, however, the population dropped drastically when the prefecture began to be raided by the "Western Plain Barbarians" (*Xiyuan man* 西原蠻). As Yuan recalls, "In the Kuimao year (763), I, the Insouciant Old Man, was appointed Prefect of Daozhou. In the past Daozhou had over forty thousand households, but now after being invaded by rebels the number is reduced to less than four thousand, and most of the remaining households are no longer able to pay taxes" (*Guimao sui, mansou shou Daozhou cishi, Daozhou jiu siwanyu hu, jingze yilai, buman siqian, daban busheng fushui* 癸卯歲，漫叟授道州刺史，道州舊四萬餘戶，經賊已來，不滿四千，大半不勝賦稅).[37]

A bleaker picture of postwar Daozhou is found in Yuan Jie's "Memorial Giving Thanks to His Majesty" ("Xieshang biao" 謝上表), which he wrote to the emperor after he was sent to Daozhou by the central government. In the memorial, he described the local scene he encountered as follows:

> When they greeted me, they groveled and cried, and the officials were pale from malnutrition. Wild grasses alone grew in the cities and villages. I ascended a height and gazed afar, but saw no trace of human habitation. In the several prefectures of Lingnan abutting my place, rebels swarmed like ants, and had not yet surrendered.
>
> 耆老見臣，俯伏而泣，官吏見臣，以無菜色。城池井邑，但生荒草，登高極望，不見人煙。嶺南數州，與臣接近，餘寇蟻聚，尚未歸降。[38]

In this passage, Yuan's prose moves rapidly from devastated men to abandoned habitations, from the absence of human traces to the disorder threatening adjacent areas. It is a symbolic survey of the territory of Daozhou by an imperial agent, and it extends the point of view to a larger geographic scope so as to illustrate the danger Daozhou faced. As is seen here, in Yuan's writing, Daouzhou is represented as an area that underwent catastrophic social and environmental changes and was pervaded by an atmosphere of desperation.

The central government sent Yuan to Daozhou in the hope that he could defend the territory from the Western Plain Barbarians. In the same memorial, Yuan expressed his will to settle the people and reconstruct the area:

> My plan is to summon and settle the refugees, set a good example to encourage the poor and the weak, defend the cities and villages, and burn the mountain forests and use the land for agriculture, with the hope that the people can all survive past the fall.
>
> 臣見招輯流亡，率勸貧弱，保守城邑，畬種山林，冀望秋後，少可全活。[39]

"Memorial Giving Thanks to His Majesty" is a generic name for the kind of formulaic document that all officials were supposed to write shortly after they started their posts. Accordingly, it was quite normal for an official to assert his future actions in a determined and optimistic way. In this case, Yuan Jie carried out his plans to a significant extent. This is demonstrated in Yan Zhengqing's tomb inscription for Yuan: "After the prefecture was destroyed by the Western Plain Barbarians, the population was reduced to less than one tenth [of its size before the unrest], and only one thousand households remained. When Lord [Yuan] took office, he carried out the ancients' way of ruling. More than ten thousand people returned to their homes and the rebels dared not invade again" (*Zhou wei Xiyuan zei suoxian, ren shiwuyi, hu cai manqian. Jun xiache, xing guren zhizheng, ernianjian guizhe wanyu jia, zei yi huaiwei, bugan laifan* 州為西原賊所陷，人十無一，戶纔滿千。君下車，行古人之政，二年間歸者萬餘家，賊亦懷畏，不敢來犯).[40]

As circumstances improved, Yuan started to engage in a reconstruction of Daozhou's landscape, including building, naming, writing, and inscribing. For him, the reconstruction of landscape involved both imagining a new ideal home and establishing authority over it. These two emphases can be seen in his essay "Record of the Cold Pavilion" ("Hanting ji" 寒亭記):

> In the year Bingwu of the Yungtai era (766), I was on an inspection tour of the districts under my jurisdiction and arrived at Jianghua. The magistrate Qu Lingwen consulted me thus: "In the south of this district, a stream and a rock reflect each other and it is a lovely sight. It was said that one could not climb up there, but I sought a way. I came across a cave and entered it and built a walkway over the dangerous sections so that I could pass. Thus, I was able to construct a thatched pavilion on top of the rock. Once the pavilion had been completed, stairs and railings were built projecting out from the rock as it looks out on the lengthy river below. The pavilion's pillars and roof mingle with the clouds and are as high as the summit. When the atmosphere is clearest at dawn and at dusk, the mist appears in unusual colors; the surrounding, blue-

> green rock wall reflects the water and the trees. I wanted to name this pavilion but could not find the words to describe it. I venture to invite you to name it for all generations to come." So I discussed the matter with him at the pavilion, saying: "Though I visited this at the height of summer, it feels as if the season will soon turn to winter. Though located in these torrid parts, it is cool and comfortable. Is it not appropriate to name it 'Cold Pavilion'?" So I wrote out this record for the Cold Pavilion and had it engraved on the rear wall.[41]
>
> 永泰丙午中，巡屬縣至江華，縣大夫瞿令問咨曰："縣南水石相映，望之可愛，相傳不可登臨。俾求之，得洞穴而入，棧險以通之，始得構茅亭於石上。及亭成也，以階檻憑空，下臨長江，軒楹雲端，上齊絕巔。若旦暮景風，煙靄異色，蒼蒼石墉，含映水木。欲名斯亭，狀類不得，敢請名之，表示來世。"於是于亭上為商之曰："今大暑登之，疑天時將寒。炎蒸之地而清涼可安，不合命之曰'寒亭。'"乃為寒亭作記，刻之亭背。[42]

As described in the essay, the county magistrate Qü Lingwen 瞿令問 (fl. 760s) kept a pavilion built in a hitherto unreachable spot. In a courteous gesture, he gave Yuan Jie, the higher-ranking official, the privilege of naming the pavilion. Qü was an accomplished calligrapher who collaborated several times with Yuan to inscribe new local landmarks.[43] His description of the view of the pavilion, highlighting the unusual color of the mist and the reflection of the water and the trees on the rock, exhibits a refined aesthetic taste. Quite surprisingly, Yuan steers away from Qü's vivid depiction of the visual pleasures of the scenery and chooses a simpler name, the "Cold Pavilion," giving the reason that the pavilion is cool and comfortable.

Qü and Yuan's contrasting imaginaries of the pavilion manifest their different approaches to landscape. If Qü speaks as a sightseer/explorer of the landscape who focuses on the visual enjoyment it may bring to the viewer, then Yuan speaks essentially as a dweller in the landscape. Yuan's strikingly straightforward naming is a reminder to both Qü and "all generations to come" that the experience of landscape should encapsulate holistic bodily sensations, and that it should make the visitor "comfortable" (*an* 安), that is, make him feel at home. He depicts the pavilion as a partial enclosure that connects the river below and the summit above, which allows the air to flow in and out to mitigate the intense summer heat. For Yuan, a former refugee from the north, a cool pavilion can shelter him from the "torrid parts" of the south and recreate surroundings that feel closer to his hometown. The name he gives to the pavilion therefore betrays his fundamental preoccupation with habitation in the south.

Moreover, Yuan's naming of the pavilion is also a declaration of his professional authority over the local magistrate. As Xiaoshan Yang observes, Yuan's response shows no appreciation of Qü's achievements as the local magistrate: "Instead, through the act of naming, Yuan Jie asserts the priority of his own brief experience of the pavilion to Qü Lingwen's more intimate and enduring relationship to the place. Thus, the two individuals' relationship to the natural landscape is subsumed in the more important hierarchical relationship between the prefect and the county magistrate." Although the text tactfully leaves out Qü's response to the new name, it is quite clear to the reader that "through the act of naming the pavilion Yuan foregrounds the assertion of his own authority and power to the detriment of his subordinate—intentionally or unintentionally."[44] As if to further consolidate his authority, Yuan inscribed his "Record of the Cold Pavilion" on the rear wall of the pavilion.

In a similar fashion, Yuan expresses his affection for a scenic spot in Daozhou as a potential permanent home in his "Inscription on the Yanghua Rock with Preface" ("Yanghua yan ming bing xu" 陽華岩銘幷序):

> Six or seven *li* south of Jianghua County in Daozhou is Hui Mountain. The southern side is steep and beautiful, and below is a huge rock facing the sun which I have accordingly named Yanghua [lit. Sunlight]. I have traveled and inhabited mountains and forests for almost thirty years, and I have never seen any place

comparable to Yanghua, with its distinctive springs and rocks, and also its suitability for building a home. So I composed an inscription to praise it. County magistrate Qü Lingwen is good at the art of ancient style inscription. Emulating the classics inscribed on stones, he carved the inscription under the rock. Here is the inscription:

Ten thousand peaks of the Jiuyi Mountain cannot compare
to Yanghua Rock.
The rock is steep and high,
Under it is a place suitable for inhabitation.
The rock is located at the mouth of the cave, facing the sun.
The rock is high and the air is clear; the cave is deep
and the spring is cool.
The rock whorls up and its summit seems broken open.
Ditches, dykes, pines, and bamboo reflect light among the water
and stones.
This place is especially good for hermits, and for retired gentlemen.
I wish to give up my official staff of command,
and live out a poor life here,
For I fear that people will criticize me
for using my power to force our times to change its course.
Making this inscription is a way to distinguish my name and deeds.
As ugly as such.
Alas! Yanghua!
Leaving, I already want to return.
I go forward but stop to gaze back, uncertain and hesitant.

道州江華縣東南六七里有回山。南面峻秀，下有大岩，岩當陽端，故以陽華命之。吾遊處山林幾三十年，所見泉石如陽華殊異而可家者，未也。故作名稱之。縣大夫瞿令問，藝兼篆籀，俾依石經，刻之岩下。銘曰：

九疑萬峯，不如陽華。陽華嶄巉，其下可家。洞開為岩，岩當陽端。岩高氣清，洞深泉寒。陽華旋回，岑巔如闢。溝塍鬆竹，輝映水石。尤宜逸民，亦宜退士。吾欲投節，窮老於此。懼人譏我，以官矯時。名跡彰顯，醜如此為。於戲陽華，將去思來。前步卻望，踟躕徘徊。[45]

This inscription demonstrates how Yuan Jie tends to appreciate landscapes from both practical and aesthetical perspectives. The Yanghua Rock is admittedly "steep and beautiful," but the decisive reason for Yuan's attraction to this place is that it offers a habitable karst cave, the Yanghua Cave. The description of the cave indicates that it sits facing the south and receives plenty of sunlight. Moreover, "the rock is high and the air is clear; the cave is deep and the spring is cool." As one experienced in finding dwelling places, the narrator is primarily concerned with the living necessities that a given place provides, including air, water, space, and a livable temperature. Only after these criteria are met does the speaker move on to describe the scenic beauty at the Yanghua Rock summit. In other words, the potentially homey cave is foundational to his evaluation of the site, while the summit serves as an outdoor open space for aesthetic appreciation. The description therefore summons an ideal image of a home with a view. Yuan then turns to fashioning the political stance implied in dwelling in such a place. He claims that the site is particularly suitable for hermits (*yimin* 逸民) and retired gentlemen (*tuishi* 退士) and that he himself would want to spend his late years there. Dying a hermit in this remote and beautiful place would indicate his integrity and fend off detractors of his ambitious political career. In the end, although the narrator leaves the place with a sense of uncertainty for his future political career, he expresses his hopes for a return.

The idea of the Yanghua Cave as an ideal dwelling place already had a history specifically related to the migration after the Rebellion. Yuan reveals this history in a poem titled "Written to Summon Administrative Aid Tao to Build His Home at Yanghua" ("Zhao Tao Biejia jia Yanghua zuo" 招陶別駕家陽華作):

> The realm within the seas has grown tired of wars.
> Twelve disturbing years have passed.
> People living in the Yanghua Cave
> Do not seem to know the chaos outside.
> Those who build homes here,

Will persevere into old age.

海內猒兵革
騷騷十二年
陽華洞中人
似不知亂焉
誰能家此地
終老可自全[46]

In this section of the longer poem, the Yanghua Cave is represented as having sheltered its inhabitants from the chaos of the Rebellion. The enclosed, isolated but well-conditioned environment of the cave would provide a sense of security for the uprooted migrants. The speaker then projects the past onto the future and asserts that it will be a perfect site to build a home.

Yuan Jie's strong identification with the Yanghua Rock as a home also reminds us of his personal experience taking refuge in a cave when he fled south from the war-torn north. We mentioned earlier that as soon as the Rebellion broke out, Yuan led more than two hundred households from his hometown into the mid-Yangtze region. They established a refugee enclave in a mountain cave in Wuchang County and lived there for months. For Yuan, the cave was both a livable and an enjoyable environment, for he was able to read and write with pleasure there.[47] The experience of living in the cave was so defining to Yuan that he styled himself as "Yiyu zi" (猗玗子; lit. the Master of Yiyu) following the name of the cave "Yiyu Cave" (*Yiyu dong* 猗玗洞).[48] As someone who named himself after a cave, it would be natural for him to be impressed by the Yanghua Cave's suitability as a site for a home. Accordingly, when he had his "Inscription on the Yanghua Rock with Preface" engraved on the wall of the cave, he embedded his personal memory of migration into the landscape itself.

At the same time, the purpose Yuan had in recording his envisioning of the Yanghua Rock as a home was not merely personal nostalgia. Rather, it was also a means to promote and develop the place. The poem "Written to

Summon Administrative Aid Tao to Build His Home at Yanghua" quoted earlier is indeed a poem to persuade his friend Tao to build his home there. In the rest of the poem, Yuan continues to detail a picture of a pleasant life in the cave to allure his friend. This is not the only case in which Yuan used his landscape writing to encourage new potential dwellers. In a poem titled "Describing Hui Creek to Summon the Retired" ("Shuo Huixi zhaotui zhe" 說洄溪招退者), he marvels repeatedly at the fertility of the creek valley, before promoting the creek banks as a desirable home site to his friends: "Now I would like to be an old man of Hui Creek, / Who will live to the east and west of my house?" (*Wujin yuzuo Huixi weng, shuineng zhuwo sheng xidong* 吾今欲作洄溪翁，誰能住我舍西東?)[49] As we can see, Yuan's writings about Yanghua Rock and other landscapes in Daozhou were not simply an expression of aesthetic appreciation combined with personal memories, but they also involved a communal effort to promote the local settlement of northern migrants. In a time when land transactions were particularly active, settling in a given place almost always started with discovering and acquiring livable land with an agreeable landscape.

Yuan Jie carried out this style of spatial configuration that regrounded the personal within the collective in this planning and construction of his own home. He had passed through a beautiful creek in Yongzhou, a prefecture adjacent to Daozhou, several times during his trip along the Xiang River. Later, he named the creek, the terrace, and the pavilion Wu Creek (*wuxi* 浯溪, lit. my creek), Wu Terrace (*wutai* 峿臺, lit. my terrace) and Wu Pavilion (*wuting* <廣+吾>廎, lit. my pavilion) and wrote three inscriptions for them.[50] The three inscriptions declared his exclusive ownership of the new landmarks, which is exceptional for Yuan. For example, the following is his "Inscription for the Creek of Mine with Preface" ("Wuxi ming you xu" 浯溪銘有序):

> This Creek of Mine lies to the south of the Xiang River and flows northward into the Xiang. As I loved its unusual scenery, I made a home in the area. This was a creek that had been nameless for

> ages. Because I love it, I have named it "This Creek of Mine" and written an inscription at the mouth of the creek.
>
> The inscription reads: "The Xiang River zigzags; its deep torrent flows along the mountain. The mountain opens up a stone gate, where the creek murmurs. How does the mountain open up? Like a pair of stones towering. Broken cliffs facing the deep, steep rocks flanking the creek. The water is extremely strange; the rocks are remarkably unusual. I wanted to seek retirement and spend my old age in this spot. This ancient creek on the waste land, deserted for a long time, I named "This Creek of Mine" to commend my exclusive possession. For those who would come to visit, I inscribed this at the mouth of the creek."[51]
>
> 浯溪在湘水之南，北匯于湘。愛其勝異，遂家溪畔。溪世無名稱者也，為自愛之，故命曰浯溪，銘于溪口。
>
> 銘曰：湘水一曲，淵洄傍山。山開石門，溪流潺潺。山開如何？巉巉雙石。臨淵斷岸，夾溪絕壁。水實殊怪，石又尤異。吾欲求退，將老茲地。溪古荒，溪蕪沒蓋久。命曰浯溪，旌吾獨有。人誰遊之，銘在溪口。[52]

This inscription is a plain announcement of his "exclusive possession" of the creek, as it is both carved in stone and embedded in the creek's name. The character *wu* 浯 combines the radical for "water" and the phonetic radical *wu* 吾. Yuan Jie uses the phonetic radical as the semantic radical for "I/my" so that the name Wu Creek 浯溪 literally means "This Creek of Mine." He does the same in naming the terrace and the pavilion. In these unusual cases, Yuan appropriates the public for the private; following him, people must call these places "my creek," "my terrace" and "my pavilion," still from Yuan's perspective.[53] By writing about this personal site and inscribing it with these invented names, Yuan turned an unclaimed landscape into part of his home.

Five years later, however, Yuan made the bold decision to inscribe a monumental piece about the fate of the empire near his home, thus spatially connecting his celebration of his private ownership of the landscape to the revival of the empire. He invited his close friend, the renowned calligrapher Yan Zhenqing, to craft his "Ode to the Revival of the Great Tang" ("Da Tang zhongxing song" 大唐中興頌), and then had the piece engraved on a cliff where the creek lets out into the Xiang River.[54] The cliff is just steps away from his earlier inscriptions and literally marks the entrance to his home if one were to approach by water. This piece was an impassioned celebration of the revival of the Tang, and its scope and imperial vision went far beyond Yuan Jie's other inscriptions in Daozhou:

> In the fourteenth year of the Heavenly Treasure era [755], An Lushan sacked Luoyang. The following year, he sacked Chang'an, the Son of Heaven graced Shu, and the heir apparent ascended the throne at Lingwu. The year after that, the emperor moved his armies to Fengxiang. That year he regained both capitals. The retired emperor returned to the capital. Alas! For the virtuous rule of the emperors of earlier dynasties, one must look to songs of praise. Now if a song of praise for [Suzong's] rule is to be engraved in stone, who would be suited to write it, if not one who has grown old in literature?
>
> Alas! In the bygone court, perverse ministers, traitorous and arrogant, caused disorder and weird portents. Frontier generals galloped their troops and brutally rebelled against the state, and the populace lost its tranquility. When the emperor made his southern inspection tour, the court officials skulked away and offered themselves to the brigands as ministers. But heaven shone on Tang and dawned on our emperor. Riding a solitary steed in the north, he stood alone with a single cry against the thousand battle flags and myriad emblazoned banners of the barbarian troops' vanguard. Then our armies marched towards the east, and the heir apparent pacified the barbarians, destroying and driving out that mob of murderers. He brought about a period of the restoration of the old rules, and achieved victory in good time. Never before had the dynasty had such a great success. After such terrible suffering, the imperial temples were again secure and the two emperors were happily united in the capital once again. The earth emerged

and the heavens opened, all terrestrial and celestial calamities were relieved, and auspicious blessings arrived in great number. The murderers and rebels were saturated with heaven's blessings, and, dead or alive, they endured their shame. Honorable rank was given the meritorious, the names of the loyalist martyrs were immortalized, and the imperial favor flowed over their sons and grandsons. The great virtue was in resurgence, like a mountain towering or the sun ascending, and a myriad of blessings were thus received. This all will make our great ruler's dignified voice and comportment known everywhere, too glorious for words. The Xiang River runs round to the east and west; Wu Creek runs straight into it, and here the stone cliffs align against the sky. Let the cliff be polished and chiseled, to engrave this song of praise therein, for thousands of years to come.[55]

天寶十四載，安祿山陷洛陽。明年，陷長安。天子幸蜀，太子即位於靈武。明年，皇帝移軍鳳翔。其年復兩京，上皇還京師。於戲！前代帝王有盛德大業者，必見于歌頌。若今歌頌大業，刻之金石，非老於文學，其誰宜為？頌曰：

噫嘻前朝，孽臣姦驕。為昏為妖，邊將騁兵。毒亂國經，羣生失寧。大駕南巡，百寮竄身，奉賊稱臣。天將昌唐，繄曉我皇，匹馬北方。獨立一呼，千麾萬旗。我卒前驅，我師其東。儲皇撫戎，蕩攘羣兇。復服指期，曾不踰時。有國無之，事有至難。宗廟再安，二聖重歡。地闢天開，蠲除祆災。瑞慶大來，兇徒逆儔。涵濡天休，死生堪羞。功勞位尊，忠烈名存，澤流子孫。盛德之興，山高日昇，萬福是膺。能令大君，聲容沄沄，不在斯文。湘江東西，中直浯溪，石崖天齊，可磨可鐫，刊此頌焉，何千萬年！[56]

Yuan Jie originally composed this piece in 762 when he served as a staff officer under the Jingnan military commissioner. The central event he celebrates here is the Tang government's recovery of the two capitals in 757, with Emperor Suzong portrayed as the main hero who called the loyalists to arms.[57] In the preface, Yuan justifies the composition, noting that the genre of the ode was used by emperors in earlier dynasties as celebration of great imperial achievements, and he himself has the credentials to author this piece since he is an experienced writer. It was not until 771 that he had it inscribed on stone and added the conclusion that delineates the spatial organization of his chosen site for the inscription: "The Xiang River runs round to the east and west; Wu Creek runs straight into it, and here the stone cliffs align against the sky. Let the cliff be polished and chiseled, to engrave this song of praise therein, for thousands of years to come." This conclusion describes the relational position of the Xiang River, the Wu creek, and the stone in a symbolic way. The Xiang River, "running round to the east and west," circumscribes a broad geographic space. The Wu Creek, cutting "straight in the middle" (*zhongzhi* 中直, lit. center and straight) of the Xiang River, symbolizes the righteousness of its owner. The cliffs, which "align against the sky," symbolizes the great undertaking of the empire. Located along the busy Xiang River, the inscribed cliff is certainly one of the most visible among all his discoveries in the local landscape. And yet, it is right next to his most private treasure, his own home. The inscription of the cliff thus connects Yuan's garden to the broader geographic and symbolic space of the empire.

Yuan's forceful way of describing, naming and inscribing the landscape has incurred harsh criticism. The Song Dynasty critic Ouyang Xiu 歐陽脩 (1007–1072) interprets it as an expression of egotism and consequently dismisses Yuan as a cheap vanity hunter. As he comments in his introduction to Yuan's "Inscription on the Yanghua Rock":

> Yuan Jie is a person who loves uncommon things. He always names the landscape he lives in himself and fears that the names are not uncommon enough. His writing has the same tone and lacks strength and energy, therefore, fails to reverberate. Men of noble character who wish to be considered immortal attain it inwardly and then show it outwardly, and this must be done in the natural way. Master Yan lay down alone in a dirty alley. No one saw his actions but his high reputation has remained over ten thousand ages. This is what is called the natural way.

> [Yuan] Jie was overhasty in seeking posthumous fame, and the strain is apparent.
>
> 元結好奇之士也，其所居山水必自名之，惟恐不奇。而其文章用意亦然，而氣力不足，故少遺韻。君子之欲著于不朽者，有諸其內而見於外者，必得於自然。顏子蕭然臥於陋巷，人莫見其所為，而名高萬世，所謂得之自然也。結之汲汲於後世之名，亦已勞矣。[58]

Ouyang notices that Yuan's style of naming and writing the landscape is uncommon, and he considers the tendency to be a manifestation of Yuan's obsession with his own fame. He then compares Yuan unfavorably with the ancient Confucian sage Yan Hui and argues that a true sage who desires fame would establish himself internally rather than externally, because his virtue will naturally speak for him and contribute to lasting fame. In Ouyang's eyes, Yuan was far from a true sage. Approaching Yuan's landscape writing as merely an embodiment of personal emotion and aspiration, Ouyang observes only superficiality and an inflated ego.

If we take into account the historical and geographical contexts of the mass migration, however, we come to a very different conclusion. The spatial imageries Yuan Jie conjured in his work clearly show that for him, defining and marking the landscape were not meant to impose his personal fame onto natural beauty, but rather to transform the desolate imperial frontier into a new home for wartime refugees and migrants, including himself. During his years in the deep south, he wrote a total of twenty-four landscape inscriptions and inscribed most of them in the landscape.[59] These inscriptions record how Yuan envisioned an ideal human space within the southern natural environment, how he renovated or planned to renovate the area, along with how that project would contribute to both personal comfort and communal well-being. His writings and inscriptions therefore mark the landscape as a space suitable for dwelling in a material way. Consequently, they need to be understood within the context of a collective effort, led in part by him as the local official, to domesticate the imperial frontiers in the south.

Yuan's landscape essays recorded the process of creating new landmarks and at the same time became part of the landmarks through inscription. Some of these landmarks have permanently changed people's perceptions and won individual sites admirable reputations in later geographic works. Take Wu Creek as an example. The literary historian Zhang Shuhui's study shows that there was no mention of the creek in *Maps and Treaties of the Prefects and Counties of the Yuanhe Reign,* which was compiled about half a century after Yuan claimed the creek; yet in the Song dynasty Wu Creek appeared in most major geographic works such as *Imperial Geography of the Taiping Xingguo Era* and *Records of Scenery on Earth.*[60] The fame of the place in turn attracted more literati to write about it or to inscribe verses near the water. According to Zhang's study, from the time Yuan had his "Ode to the Revival of the Great Tang" carved in stone until the early eighteenth century, forty-seven literary works were produced about the creek.[61] Today, there are altogether five hundred stone inscriptions scattered around the creek, so many that the place is now called the Wu Creek Stele Forest (*Wuxi beilin* 浯溪碑林). The Wu Creek Stele Forest has become a well-recorded regional and national landmark in present-day geographic works and guidebooks. Indeed, the literary geography led to further geographic changes. Between 1999 and 2000, the local government constructed a cement wall around the steles to protect them.[62] In this way, literature and geography continued to inform and inspire each other, in both the physical and the textual worlds.

Mid-Tang Landscape Essayists and Their Group Efforts

In the mid-Tang period, a group of writers carried on Yuan Jie's legacy of writing about, inhabiting, and building up the deep south. In addition to Liu Zongyuan, Han Yu and Liu Yuxi were also deeply involved in these efforts. In this section, I will show that homebuilding in the south was a collective project undertaken concurrently and consecutively across the area. The landscape essayists' spatial imaginaries, therefore, were not confined to the writers themselves but were sometimes meant to be shared among a group of collaborators. Their writings both recorded and participated in the collective effort to inhabit the deep south.

We will start in Daozhou, where Yuan Jie had paved the way for further development. In 814, Xue Bogao was sent to Daozhou by the central government to serve as prefect. The next year, Liu Yuxi was relocated to Lianzhou 連州, which is about three hundred *li* away from Daozhou. He saw himself as following in Yuan Jie's footsteps as he moved to Lianzhou: "It has been more than fifty years between the time Yuan was appointed and I was relocated. In this territory, I think of him as though he were standing right next to me" (*Qijian xiangju, wushi yunian. Fengjing huairen, qi you bijian* 其間相距，五十餘年。封境懷人，其猶比肩).[63] Liu kept in contact with Xue and they collaborated to establish a new landmark in Daozhou. As Liu's essay "Record of the Light-Holding Cave" ("Hanhui dong shu" 含輝洞述) reveals, Xue sent him a letter to describe a newly discovered karst cave and to invite Liu to write a record for it:

> One day he sent me a letter to boast about it: "I found an unusual area outside of town. It is outside the West Gate of the city, close to the Southern Mountains. There is a vaulted cave like a grand house overlooking the river. The surrounding area is like winding corridors and intricate rooms with patterns like pondweed, embroidery, engraving, and red lacquer. When clouds rise and the sunlight pours into it, strange images pile on top of each other. The water, stones, plants, and trees all look otherworldly. I'm guessing it was the abode of a Daoist master who ate rosy pink clouds and rode on the wind. After he cast off his human form and left, hundreds of thousands of years passed, and recently it was rediscovered by a peasant. So I inquired about its location and went there. I named it the 'Light-Holding Cave,' after what the poet said: 'Mountains and rivers hold pure light.' Master, you have distinguished yourself in the world as a powerful writer. Why not compose a record of it for me?"
>
> 一旦以書來夸曰："吾得異境于近郊。自城西門，並南山，俯江水，有石穹然如夏屋。其左右前後，又如回廊曲房，藻繡雕彤之象。雲生日入，怪狀迭發，水石卉木，杳非人

> 寰。意其嘗爲餐霞御氣者之所遊息，委蜕而去，不知其幾千百年。逮今得諸黃冠野夫。及詣而信，因名其地曰含輝洞，蓋詩家流所謂‘山水含清輝’者是已。吾子常以詞雄于世，盍爲我志焉？”[64]

Xue marveled at this cave that conjures up the image of a grand house. Much like Liu Zongyuan at the Rocky Stream, Xue associated many features of the cave with objects that one would ordinarily find in a house, including corridors, rooms, embroidery, engravings and red lacquer. The unique karst topography further triggered Xue's imagination, leading him to think that the cave was in fact a Daoist grotto heaven, whose owner had been gone for thousands of years. The spiritual atmosphere is accentuated by the name Xue gave the cave, "Light-Holding Cave" (*Hanhui dong* 含輝洞), which is an allusion to a famous poetic line by Xie Lingyun, "Mountains and rivers hold pure light."[65] The cave, like Xie's mountains and rivers, thus becomes a source of an aura that both embraces and enlightens the humans within it. To try to imbue the cave with even more cultural capital, Xue wrote to Liu to request an inscription.

Liu Yuxi decided to join Xue Bogao to complete the task of establishing the cave as a great landmark in Daozhou. Since he had never seen the cave, he drew from Xue's extensive travel experience and geographic proficiency to highlight the cave's uniqueness:

> Duke Xue lived beside rivers and lakes in his early years, and he visited famous mountains in his leisure time. To the east he explored the cave of Yu the Great [in Shaoxing, Zhejiang], and traveled to Siming Mountain [in modern Ningbo, Jiangxi], Juqu [in modern Jurong, Jiangsu], Jinhua Mountain [in modern Jinhua, Zhejiang], and Yangxian [in modern Yixing, Jiangsu]. To the south he crossed the Nine Rivers [in modern Jiujiang, Jiangxi], approached Mount Lu, and passed Pengli Lake [now Boyang Lake, in Jiangxi]. He was well versed in the records of all the landscapes in the world, evaluating and ranking them accurately. Yet when he found this cave he was speechless and could not describe it, proving that the cave was indeed a marvel.

> 若薛公者，少居江湖，閑遊名山，東探禹穴，曆四明、句曲、金華、陽羨，南過九江，薄匡廬，以涉彭蠡。天下山水之籍，存乎胸中，第其高下，銖兩不失。及是而口呿不能名，顧為奇，信矣！[66]

The enumeration of place-names in this passage represents both the scope of Xue Bogao's travels and Liu Yuxi's own geographic perception of southern China. The first group of places, including the cave of Yu the Great (*Yuxue* 禹穴), *Siming* 四明, *Juqu* 句曲, *Jinhua* 金華, and *Yangxian* 陽羨, are all in the lower Yangtze valley, where Liu was born and raised as a second-generation migrant. The second group of place names, including the Nine Rivers (*Jiujiang* 九江), Mount Lu (*Kuanglu* 匡廬), and Pengli Lake (*Pengli* 彭蠡), are in the middle Yangtze area. Finally, the cave in Daozhou is represented as exceeding all the other scenic sites in southern China, an area that was still unfolding itself to explorers. The essay ends with an inscription Liu wrote for the cave and reveals that the entire piece was later inscribed on the wall of the cave.

Similar projects were undertaken in Daozhou's neighboring Lianzhou as well. In 803, Han Yu was demoted to Yangshan 陽山, now in Guangdong province, a subordinate county of Lianzhou. His essay "Record of the Pavilion of Joyous Feasts" ("Yanxi ting ji" 燕喜亭記, 804), written in Yangshan, tells of a landscape exploration conducted by the Lianzhou revenue administrator Wang Zhongshu 王仲舒 (style name Hongzhong 宏中). Wang was also a northern man of letters, demoted to Lianzhou roughly at the same time as Han's relocation.[67] Reminiscent of Yuan Jie's "The Record of the Cold Pavilion," Han's essay celebrates a new landmark recently built by Wang in a neglected corner of the southern land:

> When Wang Hongzhong of Taiyuan was in Lianzhou, he would travel together with the Buddhist monks Jingchang and Yuanhui. One day, he went with these two men on a walk behind their residence. They climbed up an overgrown hill to a high point, gazed about, and discovered an unusually scenic spot. He had the tall grasses cut down, and the fine trees stood revealed; he had

the rocks removed, and a pure spring flowed forth; he had the refuse carted away and burned off the dead growth. Then he stood and observed it all. What protruded now stood forth as a hill; what receded opened up to reveal a valley. A depression became a pond; a cavity became a cave. It seemed as if ghosts, spirits, and other strange beings had come in the dark to aid them. Since then, Hongzhong and the two monks would go there in the morning and forget to return when evening fell. So a house was built to protect against the wind and rain, the winter cold and the summer heat.

When all this was completed, I requested permission to bestow names. The hill I called "The Hill of Patient Virtue," for though it was obscured in the past, it is now revealed, so it possesses the *Dao* of patience. The rocky ravine I called "The Ravine of Humble Acceptance," its waterfall, "Bestirred Egrets Falls," for the ravine signifies his virtue, and the waterfall, his appearance. The fertile valley I called "Golden Valley," and its waterfall, "Regulated Falls," for here the valley signifies his appearance, and the waterfall, his virtue. The cave I called "The Cave of Winter Residence" to indicate the season for entering it. The pond I called "The Pond of the Noble Man," for, when empty, it can store up his fine character; when filled, it can eliminate any evil qualities. The fount of the spring I called "The Spring of Heaven's Beneficence," for it issues forth from on high and bestows its bounty on those below. I summed up all these qualities by calling the house "The Pavilion of Joyous Feasts," praising him by the line in the ode in the *The Book of Poetry* that goes, "Marquis of Lu, joyous feasts." Subsequently, an elder among the people of the prefecture heard of this and accompanied me to observe it. He said, "The scenery of our prefecture is renowned throughout the world, but nothing elsewhere compares with 'Joyous Feasts'! Those who work the fields nearby encountered it, yet none recognized the value of this place. What Heaven has created and Earth safeguarded was meant to belong to such a man, was it not?"

Hongzhong came here after being demoted from director in the Ministry of Personnel. This is what he passed on his route here: from Lantian, he entered Shangluo, crossed the Xi rapids, came to the Han River, climbed Steep Mountain to gain a view of the

Square Citadel, proceeded out through Ching Gate, down the Min River, crossed over Grotto Lake, went up the Xiang River, traveled below Transverse Mountain, and crossed over the Dayu Ridge from Chen Prefecture. Among the alpine dwellings of monkeys and apes and the watery residences of fish and dragons, he has experienced the ultimate in hidden, distant, beautiful, and strange views. He should feel satiated by what he has seen and heard. Yet it appears he still is not satisfied. It is said, "The wise man delights in streams; the humane man delights in mountains." Hongzhong's virtuous character can be said to agree with what he loves. With wisdom he built this, and he dwells in it with humaneness. I know that the day is not far when he will leave here and "perform the sacred plume dance" at the Imperial Court. I then had this engraved in stone as a record.[68]

太原王弘中在連州，與學佛之人景常、元慧者遊，異日，從二人者行於其居之後丘荒之間，上高而望，得異處焉。斬茅而嘉樹列，發石而清泉激。輦糞壤，焚椔翳，卻立而視之：出者突然成丘，陷者呀然成谷，窪者為池，而缺者為洞，若有鬼神異物陰來相之。自是弘中與二人者晨往而夕忘歸焉，乃立屋以禦風雨寒暑。

既成，愈請名之。其丘曰俟德之丘，蔽於古而顯於今，有俟時之道也；其石谷曰謙受之谷，瀑曰振鷺之瀑，谷言德，瀑言容也；其土谷曰黃金之谷，瀑曰秩秩之瀑，谷言容，瀑言德也；洞曰寒居之洞，志其入時也；池曰君子之池，虛以鍾其美，盈以出其惡也；泉之源曰天澤之泉，出高而施下也；合而名之，以屋曰燕喜之亭，取《詩》所謂"魯侯燕喜"者頌也。

於是州民之聞者相與觀焉，曰："吾州之山水名天下，然而無與燕喜者比；經營於其側者相接也，而莫直其地。凡天作而地藏之以遺其人乎？"

弘中自吏部郎貶秩而來，次其道途所經，自藍田入商洛，涉淅、湍，臨漢水，升峴首以望方城。出荊門，下岷江，過洞庭，上湘水，行衡山之下。繇郴踰嶺，猿狖所家，魚龍所宮，極幽遐瓌詭之觀，宜乎其於山水饫聞而厭見也。今其意乃若不足，《傳》曰："智者樂水，仁者樂山。"弘中之德與其所好可謂協矣。智以謀之，仁以居之，吾知其去是而羽儀於天朝也不遠矣，遂刻石以記。[69]

As the beginning of the essay shows, Wang Zhongshu and his friends found this "unusual scenic spot" behind his residence, so he had the surroundings rearranged and built a pavilion-house there. In other words, the landscape and the pavilion were an extension of Wang's own dwelling place. The pavilion, "protect[ing] against the wind and rain, the winter cold and the summer heat," is perfectly habitable and provides a fresh vantage point for people to enjoy the surrounding landscape. After a generic description of Wang's renovations, Han adds a touch of mystical power to the locale: "It seemed as if ghosts, spirits, and other strange beings had come in the dark to aid them." With the help of the supernatural, the pavilion acquires its own aura.

Han Yu then names the pavilion and all the geographic features visible from the pavilion with allusions to the Confucian classics. He first conceives a long list of names for the hills and waters surrounding the pavilion, highlighting the virtue and appearance of its owner: he manifests "Patient Virtue" and "Humble Acceptance," being a "Noble Man" waiting for "Heaven's Beneficience." These names thus transform the entire natural surroundings into a paean for Wang. The name Han chooses for the pavilion, "Joyous Feasts," is in itself a celebration of royal grandeur and power, as it alludes to the feast held by the ancient King of Lu in the Spring and Autumn period. As Han further dramatizes through the voice of the local elder, the pavilion makes the surrounding scenery both visible and valuable, not only to outsiders but also for the first time to the locals. The landscape instantly becomes not only the best in Lianzhou, but the best in the world. If Yuan Jie portrays himself as a cultured, powerful owner of the landscape, Han more subtly fashions himself as an interpreter of the will of Heaven and Earth and imparts the glory to his friend Wang.

As if to substantiate the local elder's claim of the landscape's greatness in the world, Han recounts the route of Wang's travel to Lianzhou and puts the wide geographic scope in perspective. The trip starts from Lantian located in the greater capital area and moves on to Shangluo in modern southeast Shaanxi. It then continues to the Xi Rapids, the Han River, the Steep Mountain, the Ching Gate, the Min River, the Grotto Lake, the Xiang River, the Transverse Mountain, and Chen Prefecture, all lined across the mid-Yangtze area from the north to the south. It finally ends in Lingnan after crossing over the Dayu Ridge. Han Yu himself is familiar with the route he describes because of his own travels to Lingnan in his early years and his banishment to Yangshan. His enumeration of the diverse landscapes along the route gradually moves from the heartland of the empire to the "hidden, distant, beautiful, and strange views" in the deep south. Yet the literati traveler's quest for great landscapes would not be complete until he had built his own landmark as a kind of home.

Similarly, in Yongzhou Liu Zongyuan was in collaboration with other literati-officials to mark their new homes on the local land. His essay "Record of the New House of the Yongzhou Governor Wei Zhou" ("Yongzhou Wei shijun xintang ji" 永州韋使君新堂記), an inscription he wrote for his supervisor Wei Zhou's 韋宙 (fl. 810s) new house, records how a local governor's construction could change the look of the local terrain dramatically. Before the intervention of the governor, Yongzhou is described as follows:

> Yongzhou is located at the foot of the Jiuyi Mountain. The people who first measured the land built the city in the surrounding mountains. Its rocks are overshadowed by deep grasses; its springs hide in the mud. Various snakes flourish there, and raccoon-dogs and rats roam. Lush trees and unsightly trees, beautiful petals and poisonous flowers are mixed together and compete for space to grow. So the place is called a weedy wasteland.
>
> 永州實惟九疑之麓，其始度土者，環山為城。有石焉，翳于奧草；有泉焉，伏于土塗。蛇虺之所蟠，狸鼠之所游，茂樹惡木，嘉葩毒卉，亂雜而爭植，號為穢墟。[70]

This short passage dismisses the old town built by locals as nothing but a mess ridden with wild plants and vicious beasts, almost indistinguishable from the wasteland all around it. Liu Zongyuan then records a series of construction projects that Wei arranged, including "cutting down the weeds, removing the mud" and "burning and draining." Immediately, the clean and the dirty are distinguished one from another; the ugly is displaced by the pretty. Amazed by the scene, Wei went on to set up a house, from which he could appreciate the landscape:

> So he had the hall built for sightseeing and excursions. Everything was arranged according to what surrounded it, offering their advantages below the hall. Outside was the mountain range and the highland. Cliffs at the mountain foot were partially revealed in the forest. The green wilderness extended far into the deep blue sky, all merging into the gate of the watchtower.
>
> 乃作棟宇，以為觀游。凡其物類，無不合形輔勢，效伎於堂廡之下。外之連山高原，林麓之崖，間厠隱顯。邇延野綠，遠混天碧，咸會於譙門之外。[71]

Like the Pavilion of Joyous Feasts featured in Han Yu's essay, Wei Zhou's house also serves as a nexus that conjoins all of the surrounding geographic features, making them meaningful to the viewer in the house. At the end of the essay, the author summarizes the significance that landscape construction bears for the imperial agent's civilizing mission and development of the barbarian land:

> So he invited guests to visit, and then provided banquets and entertainment. Someone praised him and congratulated him, "Seeing what you have built, I understand your aspirations. You established this superb scenery along the natural contours of this place; don't you also wish to transform the people here following their customs and culture? You removed the unsightly and chose the beautiful, but don't you also wish to expel the merciless and support the humane? You cleansed the muddied water and let the clear water flow, but don't you also wish to abolish the corrupt and establish the incorruptible? You stood on high and gazed afar,

> but don't you also want to nurture every household and make them fully aware of your influence? If you do, then is this hall merely suitable for enjoying the comfort of grasses, plants, trees, soil, stones, waters, and springs, and for sightseeing among the mountains, plains, forests, and foothills? Let the followers of your principles see the things of minor worth and understand greatness. I, Zongyuan, ask permission to inscribe this piece in stone, and fix the stone on the wall, as well as to include it in my works as a model for the prefects."
>
> 已乃延客入觀，繼以宴娛。或贊且賀曰：“見公之作，知公之志。公之因土而得勝，豈不欲因俗以成化？公之擇惡而取美，豈不欲除殘而佑仁？公之蠲濁而流清，豈不欲廢貪而立廉？公之居高以望遠，豈不欲家撫而戶曉？夫然，則是堂也，豈獨草木土石水泉之適歟？山原林麓之觀歟？將使繼公之理者，視其細，知其大也。”宗元請志諸石，措諸屋漏，以為二千石楷法。[72]

The passage presents an allegorical reading of geographic features, as well as of their organization. Designing a landscape based on the nature of the local terrain is considered an indication of the governor's desire to civilize the natives without denigrating their customs; displacing the ugly and dirty in favor of the pretty and the clean reflects his determination to select moral officials to govern. More relevant to our discussion, standing high and gazing afar embodies the governor's will to bring security and prosperity to every household in the local sphere. Here, a private house becomes a vantage point from which to see a broad picture of communal living. Liu therefore praises the house for embodying exemplary governance and begs permission to have his record engraved in a stele. As we have seen in the case of Yuan Jie, Liu Yuxi, and Han Yu, Liu Zongyuan's essay combines a celebration of the reconstructed local landscape, an acknowledgement of his colleague's virtue as a dutiful local official, and an ideal for building an inhabitable space in the periphery of the empire.

Liu Zongyuan's Art of Dwelling in Yongzhou's Landscape

In this section, I will focus on Liu's ingenious portrayal of his experiences in the deep southern landscape and provide new insights into some of his most celebrated landscape essays. A more accomplished writer than Yuan Jie, Liu further developed the themes and patterns that Yuan had experimented with in his landscape essays. The locale in question is Yongzhou, traditionally part of the Cangwu area, where Yuan built his own home. Liu's landscape essays are subtle and innovative records of his search for a dwelling place. They not only capture the ongoing transformations of the local land, but also reflect his new geographic understanding of the empire.

In recent decades, scholars of Liu's landscape essays generally approach them as a result of political relocation to the periphery of the empire, which they refer to as either banishment or exile. This approach is useful in that it captures a common reality confronted by many Tang literati. As Madeline K. Spring points out, "political exile in the Tang Dynasty was a very real phenomenon, and for some writers even a temporary infatuation with the ideas and policies represented by a particular bureaucratic clique might dictate that harsh conditions be endured for the remainder of their lives. Banishment meant public and private humiliation for Tang officials, and many lived in fear of such a fate."[73] In the early ninth century, Han Yu, Liu Zongyuan, Liu Yixi, and Bai Juyi were all forced to live with exactly this fate and strove to find ways to come to terms with their situation. Among all these literary masters, Liu Zongyuan was a long-term victim of political relocation. After the defeat of the Wang Shuwen faction in 805, Liu Zongyuan, as one of the major participants, was persecuted by the court of Emperor Xianzong. As it later turned out, he was never able to reestablish his career. He was first demoted to Yongzhou, where he served nominally as an adjutant (*sima* 司馬). The post itself came with little political prestige, and Liu Zongyuan, as a demoted official, wielded even less political influence. He was not called back to the capital until a decade later, but just one month after arriving in Chang'an, he was appointed the prefect of the even more remote and less developed Liuzhou, where he wrote much less than he had in Yongzhou. He died there in 819. This brief overview shows the considerable tension between Liu and the central authorities, which had a significant influence on many aspects of his life and writing.

Understandably, many literary historians have used banishment as a key to understand Liu Zongyuan's landscape essays written in the deep south, in biographical or psychological readings of the texts. Spring's comments on the relationship between Liu and the landscape are typical. According to her, Liu's landscape essays "describe the surrounding mountains, ponds, streams, and hills that he explores as he attempts to come to terms with his plight. Frequently autobiographical, these essays show Liu both as observant naturalist and introspective philosopher."[74] Readings such as these do have a persuasive power. The inherent logic is simple but effective in working out the kind of coherent narrative commonly seen in literary history: banishment compels writers to explore the aesthetically satisfying natural environment, and at the same time this is an exploration of their intensely afflicted inner selves. In other words, literature is the payoff of an existential crisis, and the bumpy road of exile leads to the realms of wilderness, beauty, and mystical epiphany.

Thus interpreted, the landscapes of the periphery also become a metaphor for Liu's marginalized self. Along this line of thought, William Nienhauser identifies Liu's overriding metaphor as "that of spots of natural beauty isolated and neglected by their creator representing a good man who has been isolated and neglected by his friends and monarch."[75] Mason Gentzler states that the "abandoned land and the neglected beauties of nature symbolize Liu's neglected merits and talents."[76] These examples are all taken from scholarship in English. In Chinese and Japanese, a great number of similar views feature in various biographies and critical studies of Liu Zongyuan, as well as in comprehensive literary history books and encyclopedia entries.[77]

As persuasive as they may be, assessments of Liu's works within the rubric of exilic literature share a common oversight. They generally assume that the southern landscape is a static, hidden corner of nature waiting to be discovered and appreciated by Liu as the solitary landscape essayist. What is neglected in this view is the fact that the southern landscape had been a busily shifting ground ever since the An Lushan Rebellion, when generations of northern migrants negotiated with the unfamiliar living environment there. As our earlier discussion of Liu's "Discourse of the Snake Catcher" demonstrates, Liu's banishment in effect served to push him into a deep south that had already been inhabited by generations of northern migrants. A renewed close reading of his essays will show that the historical turbulence enabled him to see and write about landscapes beyond a personal horizon.

In his landscape essays, Liu Zongyuan enters into an intimate interaction with local landowners who are suffering economic crises. "Flatiron Pond" ("Humu tan ji" 鈷鉧潭記) records how Liu discovers a satisfying scenic spot and purchases the land. The place features a smooth integration of surfaces and dimensions suited to a quiet country life: "The water's froth then forms swirling wheels and slowly moves on. The calm and clear surface of the pond extends for more than ten acres. Surrounding it are trees, and there is a spring high above it"[78] (*Liumo chenglun, ranhou xuxing, qi qingerping zhe qie shimuyu. Youshu huanyan, youquan xuanyan* 流沫成輪，然後徐行，其清而平者且十畝餘。有樹環焉，有泉懸焉).[79] In other words, he deems this piece of land to be suitable as a dwelling place. He is not the first to discover the place, as it already has an owner. After Liu purchases the land, he builds his own house there:

> There was an inhabitant who dwelled above the pond, and, because I often visited it, he came knocking at my door one day and said, "I cannot pay the taxes and the debts that have piled up so I have cleared some uncultivated land in the mountains and moved there. I would like to sell off the fields above the pond for cash to settle what I owe." I gladly agreed to what he [the landowner] said. Then I increased the height of the viewing terrace, extended the

> railings, and diverted the spring to a higher place from which it could cascade down into the pond and make a splashing sound. It was especially fitting for viewing the moon during the Mid-Autumn Festival. I could then see the height of the sky and the atmosphere stretching into the distance. What else but this pond could make me glad to dwell on barbarian land and forget my longing for home?[80]
>
> 其上有居者，以予之亟游也，一旦款門來告曰：“不勝官租私券之委積，既芟山而更居，願以潭上田貿財以緩禍。”予樂而如其言。則崇其臺，延其檻，行其泉於高者而墜之潭，有聲潨然。尤與中秋觀月為宜，於以見天之高，氣之迥。孰使予樂居夷而忘故土者，非茲潭也歟？[81]

In this passage, the larger social background of land consolidation is salient. As Liu Zongyuan shows, the previous owner of the land could not afford the taxes and consequently "cleared some uncultivated land in the mountains and moved there," giving Liu the chance to procure the land. Although Liu was a lower ranking official in Yongzhou, he was not, as many of his biographers and critics have suggested, a powerless political prisoner. In fact, Liu enjoyed much leisure time in the local government and was financially stable.[82] His free time was a prerequisite for his explorations and extensive literary writing, and his wealthy status relative to deprived locals enabled him to buy the land that he desired. This land transaction between two individuals epitomizes the dynamism of the transformation of local social and economic conditions, which provided incentives for the literati to constantly explore new landscapes. As a result, the terrain itself was reshaped.[83]

With this background in mind, we may read some of Liu's famous descriptions of the landscape as representations of onsite inspections for potential homebuilding projects. Scholars have noticed that what distinguishes Liu from many other landscape writers is his participatory relationship with nature. As Nienhauser has noted, Liu's records "are not merely descriptions of a scene but records of a man visiting the scene." Nienhauser finds it reminiscent of works by Xie Lingyun, the pioneer landscape writer of the Six Dynasties, but argues that there are still fine but crucial differences between the two: "In these essays Liu Tsung-yüan himself enters into his landscape subjects. Unlike the usual portrayal of an uninhabited landscape, he climbs about, exploring, enjoying, and even beautifying his landscapes....We see him not 'at one' with nature, as Hsieh Ling-yün is said to have been, but 'at terms' with it."[84] Nienhauser takes note of Liu's intention to inhabit the landscapes he discovers. In my discussion, I will push this view further to argue that Liu's focus on these landscapes goes beyond personal aesthetic appreciation and represents the very practical purpose he and many other landscape essayists had in seeking out uninhabited areas in the deep south in the first place. What distinguishes Liu from other essayists is his ability to aestheticize the entire process of his inhabitation evaluation and to make the question of dwelling a complete sensory art.

I will use two of Liu Zongyuan's famous landscape essays to illustrate my points. The first piece is "To the Little Rock Pond West of the Little Hill." In this essay, Liu records how he abandons a unique natural area because it is unsuitable as a dwelling place. In the introductory part of this masterfully crafted essay, the narrator follows the natural music of water and makes his way into a hidden world surrounding a small pond.

> From the little hill, I walked 120 paces west and found myself blocked by a thicket of bamboo. I heard sounds of water like jingling jade pendants and bracelets and found this delightful. I had the bamboo cut down to make a path and saw below a little pond whose water was particularly clear and cold. Its bottom was entirely of rock. Along its edges, the rock bottom curved and protruded, forming rises, islets, archipelagoes, and crags. Emerald vines on viridian trees grew thickly entwined or hanging down. Uneven in length, they waved back and forth in the wind.[85]

> 從小丘西行百二十步，隔篁竹，聞水聲，如鳴佩環，心樂之。伐竹取道，下見小潭，水尤清冽。全石以爲底，近岸卷

> 石底以出，爲坻爲嶼，爲嵁爲巖。青樹翠蔓，蒙絡搖綴，參差披拂。[86]

This passage is a discovery of the sensory pleasures of nature. The list of "edges, the rock bottom...islets, archipelagoes, and crags" provides a unique nomenclature of variously shaped rocks. The diversity of the rocks contributes to the pleasant burbling of the water. The verbal sounds of *di* 坻 (rises), *yu* 嶼 (islets), *kan* 嵁 (archipelagoes), and *yan* 巖 (crags) are joined with the visual rhythm of the shared radical of "mountain" in the four terms; these sounds and images are pushed further by the repeated alliterative *wei* 爲 (to form) to produce fully orchestrated poetic music. Moreover, Liu adds to the musical scene a more visible vertical dimension consisting of the waving, hanging, and intertwined trees and vines. What follows is a vivid description of fish swimming in the clear water:

> There are a hundred or so fish in the pond who seem to be swimming in air without any support. Sunrays penetrate the water, and their shadows spread across the rock bottom as they stay contentedly still. Then suddenly they swim off, swiftly darting back and forth, as if playing with this traveler.[87]

> 潭中魚可百許頭，皆若空游無所依。日光下澈，影布石上，佁然不動，俶爾遠逝，往來翕忽，似與游者相樂。[88]

This passage has impressed many critical readers. For instance, Shen Deqian 沈德潛 praises Liu Zongyuan for his "fine techniques in depicting both the rest and the movement of the fish" (*dongding jumiao* 動定俱妙), and Chen Yan 陳衍 credits him with "being skillful at describing fish and the condition of water" (*gongyu xieyu, gongyu xieshui zhiqing* 工于寫鱼，工于寫水之情也).[89] As has been discussed in previous chapters, in terms of style and language, this passage is not so much Liu's own creation as an imitation of landscape depiction in earlier geographic works. Yet the ending of the passage, "seeming as happy as this traveler" (*si yu youzhe xiangle* 似與遊者相樂), draws the narrator into the landscape and highlights the interaction between humans and nature.

After enjoying watching the fish, the narrator shifts the tone:

> I gazed at the southwest corner of the pond, which was bent like the Dipper and wound about like a snake, the water flickering light and dark. Its edges were serrated like a dog's teeth, and I could not discover the source of the water.[90]
>
> 潭西南而望，斗折蛇行，明滅可見。其岸勢犬牙差互，不可知其源。[91]

The serpentine contours of the pond, the flickering light, the jagged edges, and the unknown water source inspire a sense of uncertainty. In the context of homebuilding, such uncertainty means insecurity. More disturbingly, Liu Zongyuan finds himself trapped in a cold, enclosed space. Although Liu mentions in the end that he was accompanied by five other people, the portrayal here is of a loner sitting by the pond, both immersed in and resisting an array of uncomfortable elements in the environment:

> I sat down above the pond and was completely surrounded by bamboo and trees on all four sides. I felt solitary without anyone else there. The scene chilled my spirit and froze my bones. I became hushed, melancholy, and remote. The scene was far too chilly to stay long, so I wrote this down and departed. Those who traveled with me were Wu Wuling, Gong Gu, and my younger brother Zongxuan. Those who came along and served us were two youths of the Cui family named Shuji and Fengyi.[92]
>
> 坐潭上，四面竹樹環合，寂寥無人，淒神寒骨，悄愴幽邃。以其境過清，不可久居，乃記之而去。同游者：吳武陵，龔古，余弟宗玄。隸而從者，崔氏二小生，曰恕己，曰奉壹。[93]

This passage has been highly praised by many critics, and this commentary by the Qing scholar Sun Cong 孫琮 (fl. 1692) is representative:

> In travel records written by ancient writers, to depict everything in a beautiful scene is not as good as not depicting it all; to explore

> every single beautiful place is not as good as not exploring them all. If everything is depicted and explored, then the taste and flavor will be drained out; if not everything is depicted and explored, then the lingering enjoyment is limitless. [In Liu's essay,] the passage about looking across to the southwest edge of the pond is an example of not depicting everything in a beautiful scene. The passage about not being able to sit very long by the pond is another example of not exploring every single beautiful place. Liu's pen and ink linger; his enjoyment has no end.
>
> 古人游記，寫盡妙景，不如不寫盡為更佳；游盡妙境，不如不游盡為更高。蓋寫盡游盡，早已境味索然；不寫盡不游盡，便見餘興無窮。篇中"遙望潭西南"一段，便是不寫盡妙景；"潭中不久坐"一段，便是不游盡妙境。筆墨悠長，情興無極。[94]

As Sun rightly points out, Liu's unique arrangement of "not depicting it all" and "not exploring it all" brings the essay a lingering charm. Instead of reading the two passages from a stylistic perspective as Sun does, I would read them as representing two unsuccessful tests for potential dwelling sites. In the first case, prior to looking towards the northwest, the narrator is impressed with the scene. As an explorer, he is first led by intriguing sounds coming from behind a thicket of bamboo. After breaking a path to the pond, his attention turns towards the intriguing shapes of the rocks under water. As he looks farther, however, he realizes that the landscape to the northwest is not ideal. In the second test, he tries to sit by the pond, but finds that it seems even less habitable, and he decides to just take note of the pond and leave. In his final assessment, the negative elements of the environment in terms of potential habitation outweigh the aesthetically pleasing aspects. Read in this light, the stylistic features that so impressed Sun Cong seem only natural: there is no need to keep exploring and recording a landscape if it has already been evaluated as unsuitable for dwelling. In this light, what differentiates Liu Zongyuan from Yuan Jie and other contemporary landscape essayists is that he not only tells stories of his successes in landscape-making but includes intriguing failures as well.

His successes can also be striking. In "Rocky Stream," Liu envisions a domestic layout that combines a place for rest, an open space, and water:

> The water here exceeds Stony Brook by threefold. Broad rocks form the bottom and extend to the banks, resembling a bed, a foundation, an open mat where foods are placed, or a threshold to an inner chamber. The water covers it evenly, its flow like embroidery, sounding like a strummed zither. I lifted up my garment and waded in barefoot. I broke off some bamboo to sweep away the dead leaves and removed the decaying trees, from which we were able to construct eighteen or nineteen folding chairs. I sat down on one and found a flowing tapestry and crashing sounds below my seat. Trees the color of kingfisher feathers and rocks with patterns like dragon scales shaded me from above.[95]
>
> 其水之大，倍石渠三之一，亘石爲底，達于兩涯。若牀若堂，若陳筵席，若限閫奧。水平布其上，流若織文，響若操琴。揭跣而往，折竹箭，掃陳葉，排腐木，可羅胡牀十八九。居之，交絡之流，觸激之音，皆在牀下；翠羽之水，龍鱗之石，均蔭其上。[96]

When Liu sees the rocks along the brook, he immediately sees the furniture one would find in a house, while the water mimics fabrics and zithers in texture and sound. In an instant, a natural landscape is thereby transformed into an imaginary home. Liu has the land cleared and constructs "eighteen or nineteen folding chairs" to fill the space. By placing the seats by the flowing water, Liu creates a homey environment that best utilizes what nature affords. As he proclaims, "Did the ancients ever enjoy anything like this? Will future visitors be able to retrace my steps?"[97] (*Guzhiren qiyou lehuci ye? Houzhi laozhe youneng zhu yuzhi jianlü ye* 古之人其有樂乎此耶？後之來者有能追予之踐履耶?)[98] The creativity in his designs did indeed go beyond ancient wisdom and would continue to inspire future generations as well.

Reading Liu Zongyuan's landscape essays in terms of the history of Chinese landscape art, the contemporary literary scholar Xiao Chi argues that Liu's essays mark an important change in the focus of the field "from mountains and waters to gardens and plants." Along these lines, Xiao interprets several of Liu's landscape essays as treatises on the art of gardening. For example, Liu's description of a given terrain can be seen as a process of "examination of the ground" (*xiangdi* 相地), a term used in Chinese landscape art to refer to the selection of land and the surrounding landscape. His essays about views seen from pavilions, on the other hand, anticipate the important principle of "borrowing a view" (*jiejing* 借景) later crystalized by Ji Cheng 計成 (1582–1642), a pioneer landscape artist in the Ming dynasty.[99] This shift of focus from the natural landscape to human constructions illustrates Liu's emphasis on the issue of human dwelling in his landscape essays.

Liu Zongyuan's particular explorations allow him to reconsider the relationship between himself and the landscape, and between humans and nature, in a refreshing way. In his essay "Little Rock Citadel" ("Xiaoshicheng shan ji" 小石城山記), he immerses himself in a philosophical reflection after finding another marvelous landscape:

> Ah! I had long doubted whether a Creator-of-Things existed or not. But now I believe that he does indeed. Yet I find it strange that he did not situate this in the Central Plains instead of placing it here in the barbaric wilds. For hundreds, for thousands of years, it has had no chance to display its talents—truly a waste of labor. If this god can be so mistaken, can it be that he does not exist after all? Someone said, "This place serves to console those worthy men who have been exiled here." Another said, "The spiritual energies of this place do not produce great men, only such natural things. Therefore, in the south of Chu, there are few talented men but many outstanding rocks." I do not believe in either of these explanations.[100]

> 噫！吾疑造物者之有無久矣。及是，愈以爲誠有。又怪其不爲之於中州，而列是夷狄，更千百年不得一售其伎，是固勞而無用。神者儻不宜如是，則其果無乎？或曰："以慰夫賢而辱於此者。"或曰："其氣之靈不爲偉人，而獨爲是物，故楚之南少人而多石。"是二者，余未信之。[101]

Here, Liu as explorer contemplates how natural wonders located in the southern wilderness could possibly relate to human beings. For him, natural objects become meaningful only when discovered by men; otherwise they represent "a waste of labor." He then raises two potential elucidations of the relationship between humans and nature. The first is a human-centered view, in which a beautiful landscape in an undeserved place is there to comfort worthy men who have been sent there unwillingly. Liu's negative response to this indicates that he understands that the landscape in Yongzhou could serve as a metaphor for his own situation, but he rejects the metaphor. The second explanation is focused on nature, viewing the southern landscape as outside the human realm. Liu rejects this explanation as well. While he does not explicitly provide his own explanation, it is clear that Liu is seeking a more productive relationship between humans and the natural world, one that is neither pitiful nor pitiless.

Liu's unstated explanation can be inferred from the landscape essays we have analyzed in this section. In these essays, men actively transform nature in their efforts to explore, enjoy, purchase, and inhabit the landscape. Liu and his colleagues' homebuilding and landscape-reshaping efforts in Yongzhou transformed karst caves, secluded wonders, and uncharted depths into various landmarks, which served as home-like refuges, far from their uncertain home in the empire's center. Liu's distinct sensitivity to nature's sensual offerings was not only an innate talent, but also a learned skill that he developed through his travels and intense engagement with the land. As a result, his landscape essays have become masterpieces of the genre.

Conclusion

Like Yuan Jie's discoveries and creations, many places Liu Zongyuan wrote about also became landmarks featured in later geographic works. In the Southern Song dynasty compilation *Records of Scenery on Earth*, an inclusive encyclopedia that introduces natural and cultural landmarks across the country, we can find records of almost all the sites that have been discussed in this chapter. For example, here is part of the entry for Yongzhou: "The land connects from Chu to Yue, and the water is a convergence of the Xiao River and the Xiang River. Traces remain of of Yuan Cishan [Yuan Jie] and the good reputation of Liu Zihou [Liu Zongyuan] is still heard" (*Dilian Chu Yue, Shuihe Xiao Xiang. Yuan Cishan zhi yiji, Liu Zihou zhi fangsheng* 地連楚越，水合瀟湘。元次山之遺蹟，柳子厚之芳聲).[102] All told, in *Records of Scenery on Earth* we find twenty-three lasting landmarks that Yuan Jie and Liu Zongyuan created or made famous. We also find entries for the Light-Holding Cave in Daozhou and the Pavilion of Joyous Feasts in Lianzhou. All of these landmarks and their associated essays have made their way into the present.

Many critics have long believed that the harsh political dynamics of the mid-Tang were the primary driving force behind the creation of favorable representations of landscapes, overshadowing other important social and environmental factors. Yet, as this chapter has demonstrated, the continuous mass migration to the deep south must also be taken into consideration, as it formed a vital backdrop for the emergence of the landscape essay genre. By representing new landmarks as a combination of both home and landscape, landscape essayists responded to the widespread social destruction and dislocation of the period. They produced inventive ways of configuring and imagining the shifting ground of the empire, opening up space in the imperial vision for invigorating new realities. Moreover, these essays led to the inclusion of these newly discovered sites in geographical encyclopedias, and thus made them an indispensable source for geographical advancement.

Notes

1. Yang, "Naming and Meaning in the Landscape Essays of Yuan Jie and Liu Zongyuan," 82.
2. According to *The Columbia History of Chinese Literature*, writings in this genre have the following characteristics: "First, they relate a personal account of a journey, whether an afternoon hike to a local scenic spot or a long journey to some remote part of the empire or even to a foreign land. Second, these works are usually structured like a diary, with entries arranged chronologically. Third, authors of travel records in China pay close attention to geography (notable landscapes, famous mountains, and so forth) and historical sites (ancient battlegrounds, well-known temples, tumuli of famous people, and so on). Lastly, *youji* almost always reflect the opinions and interpretations of the author himself regarding the places seen and visited during his journey." *The Columbia History of Chinese Literature*, ed. Victor Mair (New York: Columbia University Press, 2001), 555. This genre is also sometimes called *shanshui youji* 山水遊記, which literally means the records of mountains and waters. For a thorough study of the genealogy of the genre, see Wang, *Zhongguo gudai shanshui youji yanjiu.*
3. Ibid., 123–128.
4. See *Zhongguo wenxueshi* 中國文學史, eds. Zhang Peiheng 章培恒 and Luo Yuming 駱玉明 (Shanghai: Fudan daxue chubanshe, 1999), 201–204.
5. Lingnan was also called Lingbiao 嶺表 or Lingwai 嶺外 (both meaning "beyond the ranges"). The "ranges" refers to the Nanling ("southern ranges"), "a sinuous belt of several ranges the Chinese sometimes called the 'Five Ranges' (Wu Ling) because of the names of the five sections of the range," which "divides the Yangtze valley from Lingnan in a generally east-west alignment between latitude 25 and 26 N." See Robert B. Marks, *Tigers, Rice, Silk, and Silt: Environment and Economy* (Cambridge: University of Cambridge Press, 1998), 23.
6. Seamon, "Reconciling Old and New Worlds: The Dwelling–Journey Relationship as Portrayed in Vilhelm Moberg's 'Emigrant' Novels," in David Seamon and Robert Mugerauer eds., *Dwelling, Place and Environment: Towards a Phenomenology of Person and World* (Boston: Martinus Nijhoff Publishers, 1985), 227.
7. Ibid., 227–228.

8. Nicolas Tackett, *The Destruction of the Medieval Chinese Aristocracy* (Cambridge MA: Harvard University Asia Center, 2014), 99.
9. Madeline K. Spring's comment on the relationship between Liu and the landscape is typical. According to her, Liu's landscape essays "describe the surrounding mountains, ponds, streams, and hills that he explores as he attempts to come to terms with his plight. Frequently autobiographical, these essays show Liu both as observant naturalist and introspective philosopher." Spring, "Tang Landscapes of Exile," *Journal of the American Oriental Society*, vol. 117, no. 2 (Apr. – Jun., 1997), 314.
10. Along these lines, William Nienhauser identifies Liu Zongyuan's overriding metaphor as "that of spots of natural beauty isolated and neglected by their creator representing a good man who has been isolated and neglected by his friends and monarch." Nienhauser, "Landscape Essays," 75.
11. For a general study of these writers' banishment, see Shang Yongliang 尚永亮, *Bianzhe wenhua yu bianzhe wenxue: yi zhong Tang Yuan he wuda shiren zhibian jiqi chuangzuo wei zhongxin* 貶謫文化與貶謫文學：以中唐元和五大詩人之貶及其創作为中心 (Lanzhou: Lanzhou daxue chubanshe, 2003).
12. See Wu Songdi 吳松弟. *Zhongguo yimin shi* 中國移民史 (Fuzhou: Fujian renmin chubanshe, 1997), vol. 3, 234.
13. For detailed accounts of migration during and after the An Lushan Rebellion, see ibid., 234–251.
14. *Jiu Tang shu*, 39.1552.
15. See *Tang hui yao*, 68.1192.
16. Yuan lived in Daozhou for eight years from 764 to 782, with the exception of a one-year military post in Rongzhou (in now Guangxi Province). He died on his way to Chang'an to take up a post in the central government. Over the course of his life, he spent the second longest stretch of time in Daozhou; the longest was in his hometown Lushan (in now Henan province). For more about the biography of Yuan Jie, see Yang Chengzu 楊承祖, *Yuan Jie yanjiu* 元結研究 (Taibei: Guoli bianyiguan, 2002).
17. Chen, *Liu Tsung-yüan and Intellectual Change in T'ang China, 773–819* (New York: Cambridge University Press, 1992), 42.
18. See Shi, *Liu Zongyuan nianpu*, 10–11.
19. Liu, *Liu Zongyuan ji jiaozhu*, 42.2812.
20. See Liu Guoying *Han Yu pingzhuan*, 13–16. See also Charles Hartman, *Han Yu and the Tang Search for Unity* (Princeton, New Jersey: Princeton University Press), 21–22.

21. See *The Cambridge History of China, Vol. 3: Sui and Tang China, 589–906 AD, Part 1*, 24–27.
22. See Zhang Jinguang 張晉光, *Anshi zhi luan dui Tangdai jingji fazhan yingxiang yanjiu* 安史之亂對唐代經濟發展影響研究 (Beijing: Zhongguo caizheng jingji chubanshe, 2008), 113–116.
23. Ibid., 103.
24. See Liu Zongyuan, "Preface to Seeing off My Younger Cousin Mou Returning to Jiangling" 送從弟謀歸江陵序, *Liu Zongyuan ji jiaozhu*, 24.1586.
25. *Quan Tang shi*, 272.3060.
26. Chen, *Images and Ideas in Chinese Classical Prose: Studies of Four Masters* (Stanford: Stanford University Press, 1988), 102.
27. Liu, *Liu Zongyuan ji jiaozhu*, 16.1117.
28. Chen, *Images and Ideas in Chinese Classical Prose: Studies of Four Masters*, 101.
29. Liu, *Liu Zongyuan ji jiaozhu*, 16.1116.
30. According to Xiaoshan Yang, "That Yuan Jie anticipated Liu Zongyuan has been the critical consensus ever since the Song period." See Yang, "Naming and Meaning," 82. This view still holds true for many modern literary historians. Commenting on Yuan's "The Record of Right-Side Stream" 右溪記, Nienhauser says, "Although the style is much simpler and the tone less complex than in Liu's essays, nevertheless this piece is very similar in form to Liu's. Thus, it should not be claimed that Liu Zongyuan was the first writer to compose landscape essays (or record of excursions)." See Nienhauser, "Landscape Essays," 69. In his comprehensive study of landscape essays in the Chinese literary tradition, Wang Liqun credits Liu as the writer who fully established landscape essay (*shanshui youji*) as a literary genre, and considers Yuan Jie to be his forerunner. See Wang, *Zhongguo gudai shanshui youji yanjiu*, 124–128.
31. See Jiang Peng 江鵬, "Cangwu rushi kao" 蒼梧入詩考, *Wuzhou xueyuan xuebao* 梧州學院學報, vol. 2, no. 24 (2014.2): 48–55
32. *Quan Tang shi*, 184.1886.
33. See Zhou Xunchu 周勛初, *Li Bai pingzhuan* 李白評傳 (Nanjing: Nanjing daxue chubanshe, 2005), 132.
34. See Herman Ooms, *Imperial Politics and Symbolics in Ancient Japan: The Tenmu Dynasty, 650–800* (Honolulu: University of Hawai'i Press, 2009), 283.
35. *Quan Tang shi*, 160.1640.

36. See Zhou Zhenhe 周振鶴, *Xuela yishijiu* 學臘一十九 (Jinan: Shandong jiaoyu chubanshe, 1999), 130.
37. Yuan, "Ballad of Chongling" 舂陵行, *Yuan Cishan ji*, 4.1a.
38. Ibid., 10.12a.
39. *Yuan Cishan ji*, 10.12a.
40. Yan Zhenqing, "Inscription to the Tomb of Lord Yuan, the Tang's Late Commander of Rongzhou Concurrent Vice Censor in Chief and Military Commander of the Region (with a Preface)" 唐故容州都督兼御史中丞本管經略使元君表墓碑銘並序, *Quan Tang wen*, 344.3495.
41. Yang, "Naming and Meaning," 87.
42. Yuan, *Yuan Cishan ji*, 9.7a–b.
43. For example, the "Inscription on the Yanghua Rock with Preface" to be discussed later in the chapter was inscribed on the Yanghua rock by Qü Lingwen.
44. Yang, "Naming and Meaning," 87.
45. Yuan, *Yuan Cishan ji*, 6.5a–b.
46. Ibid., 4.3b.
47. Yang, *Yuan Jie yanjiu*, 41–44.
48. Yiyu is probably a homophone of *yiyu* 猗歟, which is often used as an exclamation of surprise or admiration, i.e. "Ah, yes!" or "O, indeed!" See Paul W. Kroll ed., *A Student's Dictionary of Classical and Medieval Chinese* (Leiden: Brill, 2014), 541.
49. Yang, *Yuan Cishan ji*, 4.6a.
50. Ibid., 6.6b.
51. Yang, "Naming and Meaning," 89.
52. Yuan, *Yuan Cishan ji*, 6.6b.
53. It is notable, however, that he made his home not in Daozhou, his own administrative district, but in Yongzhou. The literary historian Yang Chengzu comments that two reasons might account for his choice. One was to avert the kind of aspersions that often occurred when an official purchased land in his own administrative district; the other is that Wuxi was closer to the regional military capital than Daozhou, and was therefore more secure. In other words, Yuan's very act of private ownership may have involved some public considerations in the first place. See Yang, *Yuan Jie yanjiu*, 171.
54. Yang, *Yuan Jie yanjiu*, 183.
55. Translation by McNair with modifications, see *The Upright Brush*, 51–52.
56. Yuan, *Yuan Cishan ji*, 6.1a–b.

57. Yang Chengzu believes that in writing the ode Yuan embedded an implicit criticism of Suzong (r. 756–762) for putting his father Xuanzong (r. 712–756) under house arrest. See Yang, *Yuan Jie yanjiu*, 106–110. This may be the case, because some phrases in the ode do seem ambiguous. For instance, "the two emperors were happily united in the capital once again" may suggest that the two emperors in fact had had a problematic relationship before their reunion. In reality, even after this reunion any formal tribute paid to the retired Emperor Xuanzong was considered suspect by Emperor Suzong's court. Yan Zhenqing, the calligrapher for Yuan's inscription, was demoted for leading a group of officials who publically greeted Emperor Xuanzong.
58. Ouyang, Ouyang Xiu quanjiu 歐陽修全集 (Xianggan: Guangzhi shuju, 1970), 6.33.
59. Peng Min 彭敏, "Yuan Jie 'Cishan mingxu' chutan" 元結《次山銘敘》初探, *Hunan keji xueyuan xuebao* 湖南科技學院學報, vol. 32, no. 5 (2011.5), 31.
60. Zhang Shuhui 張蜀蕙, "Shui zai dijing shang xiezi—you 'Datang zhongxingsong' bei tanjiu Songdai dizhi shuxie de mingke yu duihua" 誰在地景上寫字—由〈大唐中興頌〉碑探究宋代地誌書寫的銘刻與對話, *Shida xuebao yuyan yu wenxue lei* 師大學報語言與文學類, vol.55, no.2, 38–45.
61. Ibid., 33–37.
62. He Jie 何捷, "1200 nian qian guren liuzai daoxian de moya shike, shenyun zhijin meiyou diushi" 1200年前古人留在道县的摩崖石刻，神韵至今没有丢失, *Chenbao zhoukan*, 晨報周刊, vol. 352 (2015).
63. Liu, "Record of the Pavilion of a Reclusive Official" 吏隱亭述, in *Liu Yuxi quanji biannian jiaozhu*, 15.1006.
64. Liu, *Liu Yuxi quanji biannian jiaozhu*, 15.1007–1008.
65. Xie, "Written in a Stone Study Surrounded by the Lake" 石壁精舍還湖中作, *Xie Kangle shizhu* 謝康樂詩注 (Beijing: Renmin wenxue chubanshe, 1958), 3.63.
66. Liu, *Liu Yuxi quanji biannian jiaozhu*, 15.1008.
67. Richard E. Strassberg, *Inscribed Landscapes: Travel Writing from Imperial China* (Berkeley: University of California Press, 1994), 458.
68. Strassberg, *Inscribed Landscapes: Travel Writing from Imperial China*, 122–123.
69. Han, *Han Yu wenji huijiao jianzhu*, 3.335–336.
70. Liu, *Liu Zongyuan ji jiaozhu*, 27.1805.
71. Ibid., 27.1805.

72. *Liu Zongyuan ji jiaozhu*, 27.1805–1806.
73. Spring, "Tang Landscapes of Exile," *Journal of the American Oriental Society*, vol. 117, no. 2 (Apr.–Jun., 1997), 312.
74. Spring, "Tang Landscapes of Exile," 314.
75. Nienhauser, "Landscape Essays," 75.
76. Gentzler, "A Literary Biography of Liu Tsung-yüan, 773–819" (Ph.D. Diss. Columbia University, 1966), 258.
77. To name a few examples: Gao Haifu 高海夫 *Liu Zongyuan sanlun* 柳宗元散論 (Xi'an: Shanxi renmin chubanshe, 1985), 157–169; Wu Xiaolin 吳小林, *Liu Zongyuan de sanwen yishu* 柳宗元的散文藝術 (Taiyuan: Shanxi renmin chubanshe, 1989), 128–132; Gong Yulan, 龔玉蘭 *Bianzhe shiqi de Liu Zongyuan yanjiu* 貶謫時期的柳宗元研究 (Nanjing: Fenghuang chubanshe, 2010), 132–135; *Liu Zongyuan da cidian*, 90. Shimizu Shigeru 清水茂, "Yū Sōgen no seikatsu taiken to sono sansuiki" 柳宗元の生活体験とその山水記, in *Chūgoku bungakuhō* 中國文學報 (1955-04): 45–74; Shimosada Masahiro下定雅弘, *Ryū Sōgen: gyakkyō o ikinuita utsukushiki tamashii* 柳宗元: 逆境を生きぬいた美しき魂 (Tokyo: Bensei Shuppan, 2009).
78. Strassberg, *Inscribed Landscapes: Travel Writing from Imperial China*, 142.
79. Liu, *Liu Zongyuan ji jiaozhu*, 29.1899.
80. Strassberg, *Inscribed Landscapes: Travel Writing from Imperial China*, 142–143.
81. Liu, *Liu Zongyuan ji jiaozhu*, 29.1899.
82. See Chen Songbai 陳松柏, "Ming wei sima, shiwei qiutu bian" "名為司馬, 實如囚徒" 辩, *Guangdong jishu shifan xueyuan xuebao (shehui kexue)* 廣東技術師範學院學報 (社會科學) (2009.1), 17–20. See also Chen, *Images and Ideas*, 72.
83. In "The Little Hill West of Flatiron Pond," Liu also congratulates himself on a successful land purchase deal: "I inquired about its owner and was told, 'This is abandoned land belonging to the Tang family. They have been trying to sell it but have not been able to.' I asked the price and he said, 'Only four hundred.' I liked it and bought it. Li Shenyuan and Yuan Keji, who had accompanied me, were both overjoyed and surprised by the unexpectedly low price. If this hill with its scenic beauty were moved to such areas as Feng, Hao, Hu, or Du, then wealthy travelers would compete to purchase it. They might increase their offers by a thousand pieces of gold each day yet still be unable to obtain it. But now, it stands neglected in this prefecture. Farmers and fishermen pass by and think it worthless. Offered at a price of four

hundred, it remained unsold for years. Shenyuan, Keji and I alone have enjoyed it and obtained it, so has it not truly found good fortune at last?" 問其主，曰："唐氏之棄地，貨而不售。"問其價，曰："止四百。"余憐而售之。李深源、元克已時同游，皆大喜，出自意外...以茲丘之勝，致之灃、鎬、鄠、杜，則貴游之士爭買者，日增千金而愈不可得。今棄是州也，農夫漁父過而陋之，賈四百，連歲不能售。而我與深源、克已獨喜得之，是其果有遭乎！See Strassberg, *Inscribed Landscapes: Travel Writing from Imperial China*, 143. Liu, *Liu Zongyuan ji jiaozhu*, 29.1904–1905.

84. Nienhauser, "Landscape Essays," 70; 79.
85. Strassberg, *Inscribed Landscapes: Travel Writing from Imperial China*, 144.
86. Liu, *Liu Zongyuan ji jiaozhu*, 29.1912.
87. Strassberg, *Inscribed Landscapes: Travel Writing from Imperial China*, 144.
88. Liu, *Liu Zongyuan ji jiaozhu*, 29.1912.
89. See Shen, *Tangsong badajia duben* 唐宋八大家讀本 (Hefei: Anhui wenyi chubanshe, 1998), vol. 9, 285; For Chen Yan's commentary, see Liu, *Liuzongyuan ji jiaozhu*, 1916.
90. Strassberg, *Inscribed Landscapes: Travel Writing from Imperial China*, 144.
91. Liu, *Liu Zongyuan ji jiaozhu*, 29.1912.
92. Strassberg, *Inscribed Landscapes: Travel Writing from Imperial China*, 144.
93. Liu, *Liu Zongyuan ji jiaozhu*, 29.1912.
94. Ibid., 29.1915.
95. Based on Strassberg, *Inscribed Landscapes: Travel Writing from Imperial China*, 146.
96. Liu, *Liu Zongyuan ji jiaozhu*, 29.1930.
97. Strassberg, *Inscribed Landscapes: Travel Writing from Imperial China*, 146.
98. Liu, *Liu Zongyuan ji jiaozhu*, 29.1930.
99. Xiao Chi 蕭馳, "Cong shanshui dao shui shi: Yuan Jie, Liu Zongyuan yu zhong Tang shanshui meigan huayu de yizhong bianhua" 從"山水"到"水石"： 元結、柳宗元與中唐山水美感話語的一種變化, *Zhongzheng hanxue yanjiu* 中正漢學研究 (2014.12), 293–332.
100. Strassberg, *Inscribed Landscapes: Travel Writing from Imperial China*, 147.
101. Liu, *Liu Zongyuan ji jiaozhu*, 29.1934.
102. Wang Xiangzhi 王象之, *Yudi jisheng* 輿地紀勝 (Hangzhou: Zhejiang guji chubanshe, 2012), 5.1535. *Wonders of the World* also has entries for the "Light-Holding Cave" in Daozhou and the "Pavilion of Joyous Feasts" in Lianzhou. See *Yudi jisheng*, 5.1570, 7. 2271.

Chapter 5

Bai Juyi and Yuan Zhen

The Wonder of Interactive Geography

This final chapter explores the intimate intersection between literature and geography in the realm of interpersonal communication in the mid-Tang. In earlier chapters, we have discussed various methods of literary engagement with geography. In this chapter, these methods are all woven into interactions between people, in which the friendship between literati and their attachment to the literary community are expressed in a spatialized manner: routes, distance, and places all become articulations of both personal and interpersonal relations and emotions. Through the literati's reorganization and re-association of a variety of spatial elements in their literary work, friends residing in different regions could be united in a textual space, and the destinations to which their texts travel are connected into a vision of a collaborative empire of poetry. This involved evocations of a mental map, or a navigational space envisioned in the mind's eye, in which one can trace the motion of a person or a text through one's own imagination. The chapter highlights the literary authors' agency that rendered geography a register of literati bonding and literary flourishing.

The central figures of this chapter are two cultural celebrities of the mid-Tang literary world: Yuan Zhen and Bai Juyi. Often separated by substantial geographical distances, they nonetheless maintained a faithful relationship of mutual trust and admiration throughout their eventful lives, and they transformed this into literary articulations that delighted a wide range of readers across the empire. More central to our discussion, however, are the extraordinary ways the two poets mobilized their literary talent and geographic expertise to create a shared textual space in defiance of the physical expanse of the empire. There they interacted seamlessly, sometimes in surprising synchrony, sometimes in artful comparison, and sometimes in dramatic confluence. Thus, their joint literary pursuit is a testament to the versatile strength of the mid-Tang spatial-poetic imagination.

Bai Juyi has the largest collection of extant poems of all the Tang poets. His masterpieces like the "Song of Enduring Sorrow" ("Changhen ge" 長恨歌) and the "Ballad of the Pipa" ("Pipa xing" 琵琶行) are among the most popular poems in the history of Chinese literature. Yuan Zhen does not enjoy the same high fame today, but as an innovative literary mind, he was widely considered Bai's equal in the mid-Tang. In addition to poetry writing, he also authored "The Story of Yingying" ("Yingying zhuan" 鶯鶯傳), which is commonly considered one of the finest of the Tang tales.[1] During their lives, Bai and Yuan developed a long creative partnership and legendary friendship that are incomparable in Chinese literary history. They exchanged a large number of poems, and saw their names bound together as "Yuan-Bai."[2]

Numerous scholars and critics have written about the literary accomplishments of the two poets, as individuals and as a pair.[3] Bai was one of the three best-studied Tang writers in twentieth-century Chinese scholarship, the other two being the High Tang poets Li Bai and Du Fu.[4] Studies of Yuan Zhen, although dwindling as Yuan's reputation declined over time, have nevertheless covered almost all aspects of his life and writing, including his political and literary activities, his various stylistic experiments in different genres, and the dissemination and reception of his work. In particular, much focus has been given to the poets' intensive and much romanticized literary collaboration. Scholars, ancient and modern, have repeatedly investigated, canonized, and mystified this relationship. To date, the image of Yuan-Bai fashioned by their writings and performances has stayed in the intellectual imagination as a perfect embodiment of the ancient ideal of *zhiyin* 知音, meaning soulmates who genuinely appreciate each other's "music."[5] The thus far neglected geographic dimension of their joint literary production, however, will add yet another vital layer to our understanding of the Yuan-Bai cultural complex.

I term the geographic elements involved in the making of the Yuan-Bai literary alliance "interactive geography." In their poetic exchanges, the two poets co-deployed a host of literary styles, tropes, and techniques that derived their expressive power from geographically informed spatial imaginaries. These imaginaries were either summoned from the geographical knowledge that the poets read or heard or conjured up as they experienced the expanse of the empire on horseback. In any case, common to these imaginaries were the juxtaposition, comparison, and association of multiple places in the poets' collaborative textual space. As they traveled and wrote, the poets came to fashion their individual and collective lives and identities in spatially conceived terms, alluded to their mutual affinity in richly informative and interactive geographic depictions, and cocreated a virtual poetic map of the transregional diffusion of their common literary fame. Their ways of thinking were vividly spatial, their means of expression conspicuously geographical. Ultimately, the two poets selected and mobilized resources from the rapidly expanding field of geography for their literary composition, so as to give the interactions between the poets and their communication with the larger literary community a compelling textual form.

In the following I will first investigate Yuan and Bai's individual works in light of their respective geographical understanding and experience, as a prelude and a reference point for the ensuing discussion on their corresponding writing. I will then move on to analyze the two poems Yuan and Bai wrote to each other in 809, while Yuan was on his way to Sichuan and Bai was in the capital. The two poems, titled "Dream in Liangzhou" ("Liangzhou meng" 梁州夢) and "Drunk and Missing Yuan the Ninth with Li the Eleventh" ("Tong Li shiyi zuiyi Yuan jiu" 同李十一醉憶元九), both feature the locations where they resided and startlingly match the timing of the two poets' movements through their own spaces. The geographical synchronization in both poems bespeaks their will and ability to trace each other through a mental map, and their ambition to address the larger literary community through their exchange poetry. A more advanced and sophisticated instance of their interactive geography is found in the two long exchange poems they wrote in the mid-810s, when Yuan was in Tongzhou in what is now Sichuan Province and Bai was in Jiangzhou in the lower Yangtze River area. They are "Song of the Southeast in One Hundred Rhymes, Sent to Attendant Censor Yuan the Ninth in Tongzhou, Secretary Li the Eleventh in Lizhou, Officer Cui the Twenty-Second in Guozhou, Supernumerary Wei in Kaizhou, Rectifier of Omissions Yu the Thirty-Second, Reminder Du the Fourteenth, Supernumerary Tutor Li the Twentieth, and Editor Dou the Seventh" (henceforth abbreviated to as "Song of the Southeast"; ("Dongnan xing yibai yun ji Tongzhou Yuan jiu shiyu Lizhou Li shiyi sheren Guozhou Cui ershier shijun Kaizhou Wei da yuanwai Yu sanshier buque Du shisi shiyi Li ershi zhujiao yuanwai Dou qi jiaozhu" 東南行一百韻寄通州元九侍御澧州李十一舍人果州崔二十二使君開州韋大員外庾三十二補闕杜十四拾遺李二十助教員外竇七校書) by Bai Juyi, and "In Response to Letian's Song of the Southeast in One Hundred Rhymes" ("Chou Letian dongnan xing shi yibai yun" 酬樂天東南行詩一百韻; henceforth abbreviated as "In Response to Song of the Southeast") by Yuan Zhen.[6] In these two extremely well known but understudied poems, a lengthy depiction of Jiangzhou by Bai is matched by an equally lengthy elaboration on both Jiangzhou and Tongzhou by Yuan. In Yuan's poem, he addresses Jiangzhou in odd lines and his own place in even lines while matching the rhyme word to the original, thus interlocking the two otherwise unrelated southern peripheries in Yuan and Bai's poetic space. Finally, I will look into how Yuan and Bai traced, altered, and dramatized the traveling routes of their texts through their corresponding literary writing and copying, thus outlining a map of their literary empire that stretches beyond the boundaries of the real and testifies to their unprecedented literary popularity.

Spatial-geographical Techniques in Yuan and Bai's Individual Poetry

Spatial-geographical techniques were an integral part of Yuan and Bai's respective poetic arsenals. Personal experiences with travel and reading geographies gave their poetry a distinctive spatial depth, subtly molding the ways they portrayed their lives and journeys. In this section, I offer a close reading of several poems the two authors composed as they ventured across the Tang imperial space. They were introduced to the imperial geography in different ways, and consequently employed different techniques to spatialize their personal experiences in poetry. The works to be discussed in this section differ from their collaborative writing in that they are freestanding pieces rather than part of an exchange. They are nevertheless significant for my analysis, because the geographic perspectives and spatial techniques they showcase constitute the building blocks of Yuan-Bai's more sophisticated and intricate engagement with geography in the context of their exchange poetry.

Let's first look at Yuan Zhen, an experienced reader of geography and practitioner of cartography. Before the age of thirty, Yuan had never traveled far from the northern heartland, and his geographical knowledge most likely came from broad textual study. In 809, an official trip to Sichuan gave him the opportunity to undertake his first adventure in the wider geographic world. As he moved further away from the familiar surroundings of Chang'an, Yuan encountered the rivers, hills, and mountain ranges that he had read about before. These were exciting moments of revelation, and Yuan records one of them in his short poem "Jialing River" ("Jialing jiang" 嘉陵江):

> As a man of the Qin I am only familiar with Qin waters.
> Yet I have always longed for the River Wu and the River Shu.
> Today in the chamber of the Jiachuan post,
> How lovely is the silk-like river circling beyond the bright window.

> 秦人唯識秦中水
> 長想吳江與蜀江
> 今日嘉川驛樓下
> 可憐如練繞明窗[7]

In the first two lines, the speaker identifies himself as a man of the ancient land of the Qin, to which his real-life geographic experience is limited. Although "longing for" the rivers of the Wu and the Shu, he has never seen them except in maps and geographic texts. The contrast between his broad geographic studies and meager geographic exposure betrays the speaker's curiosity, and indeed anxiety. The third line shifts from the long perspective of the past to the present. In that moment, the fantasies of an armchair geographer come true: the Yangtze River flows right beneath his window, like a piece of silk he could feel with his own hand. Remote names on the map become a pure pleasure in immediate sight, triggering a sense of satisfaction. This is how Yuan Zhen experienced the world as he traveled to the distant posthouse in Sichuan, as expressed in the poem. Where others might feel loneliness and fear on a first-time long-distance journey, Yuan took delight in seeing the geographic features that had intrigued him from his home in the comfort of the capital.

There is more to the poem than an expression of personal pleasure. The deliberate juxtaposition of the land of Qin with the rivers of Wu and Shu should remind us of Du Fu's line "rivers connected Wu and Shu" in his poem about a map-painting, which I analyzed in chapter 2. In Du's poem, rivers overcome cultural barriers to bring together the historically divided lands of the Wu and Shu kingdoms. Here in Yuan Zhen's poetic contemplation, the summoning of rivers all across the realm serves a more personal purpose. It creates a sense of connectivity among the various constituencies of the empire, thus building the visual effects of a traveler moving in an organically linked geographic space that was at once unexplored and intimately familiar. Geography found its way into Yuan's poetic creation, mitigating his uneasiness on the road while giving his personal feelings a spatialized expression.

As Yuan Zhen went further on his trip, a gentle yearning for the presence of his friends replaced his initial curiosity about unfamiliar lands. With a geographer's apt eye, he maps out his personal fondness in indirect spatial terms in "The Luokou Posthouse" ("Luokou yi" 駱口驿, 809):

> Two stars serve at the frontier of southern barbarians.
> Overnight by lamplight a Hanlin scholar drafts an imperial edict.
> In Dongchuan, I am exactly halfway between them.
> I see a moon in the south and clouds in the north.

> 二星徼外通蠻服
> 五夜燈前草御文
> 我到東川恰相半
> 向南看月北看雲[8]

This is a poem born of happy coincidence. On Yuan's way to Sichuan, on the walls of a posthouse at Luokou, he saw poems written several years before by Bai Juyi and two other acquaintances, Li Fengji 李逢吉 (758–835) and Cui Shao 崔韶 (dates unknown), and decided to add his own poem. Opting against a direct outpouring of his innermost emotions, the author calmly visualizes his friends as they worked late into the night in places scattered across the realm. Li Fengji and Cui Shao, the two "stars," are probably still studying frontier policies in their posts in the far south. Bai Juyi, Yuan's best friend, must be drafting an edict for the emperor under flickering lamplight in the capital. Finding himself halfway between the two remote places, Yuan takes momentary joy in looking up to the same bright moon hovering over the southern sky and the same clouds covering Chang'an.

Throughout this short poem, Yuan conveys his personal longing in oblique geographic terms, using the spatial distance between himself and his friends as a metaphor for his deep, inexpressible feelings. At the same time, however, the poetic technique also conjures up a strong sense of simultaneity through its montage-style juxtaposition of disparate spaces in the same scene. Companions everywhere in the imperial realm are working at the same pace in this nocturnal moment. This thought at once imparts a sense of community to the vivid poetic map and transforms personal affinity into dynastic loyalty on a larger level. The posthouse on the road thus acquires greater significance as a point of connection that not only links individual members of the literary community, but also joins the center, a semi-periphery, and the southern periphery of the empire.

In 818, Yuan Zhen again combined his sharp geographic consciousness with spatialized literary techniques, staging a show of affection for Bai Juyi. The poem, titled "A Poem Sent to Letian [Bai Juyi], Composed on the River" ("Shuishang ji Letian" 水上寄樂天), was written in Xingyuan (present-day Hanzhong in Shaanxi Province) and later sent to Bai Juyi, who was demoted to Jiangzhou (in present-day Jiangxi Province). It embodies a new-found harmony between geographic imagination and innovations in the poetic style, foreshadowing Yuan's later literary exchanges with Bai:

> The moonlit water in front of my eyes
> First flows into the Han River
> The Han River flows to join the Yangtze River and then the sea
> The Western River passes by the Yu Tower.
> The Yu Tower has a moon over it tonight,
> But are you standing on the tower?
> If there is a chance that you are gazing afar from the tower,
> I hope you can see my sorrow.

> 眼前明月水
> 先入漢江流
> 漢水流江海

西江過庾樓
庾樓今夜月
君豈在樓頭
萬一樓頭望
還應望我愁[9]

Yuan Zhen organizes several place names into the flow of poetic lines, showing intimate knowledge of the geographic features between Xingyuan and Jiangzhou. Unfolding in front of the reader is a large-scale vision of the Yangtze River system: the rivers of the Xingyuan region flow into the Han River, which then joins the Western River. The Western River traverses Jiangxi before meeting the Eastern Sea. Clearly the constant flow of water is a metaphor for the poet's deep yearning for his friend, whom he hopes is also standing by the river, gazing in his direction across all the waves and fields that separate them.

As if to give further force to the unrelenting waves of water, the poet arranges the lines in the form of anadiplosis, or a series of relayed repetitions of the characters from the preceding lines. The character relay in the form mimics the relay of river streams in the geographic imagination, pushing the reader's vision ever closer to where Bai resided in a powerful synergy between the literary and the geographic. This stylistic device is very much of Yuan's own creative invention. As the Qing Dynasty critic Wu Qian 吳騫 (1733–1813) points out, "The structure of the poem has been rarely seen in history " (*Cige gujin jueshao* 此格古今絕少).[10] It is even rarer to use it to embody both a sense of regional continuity and a personal connection with friends. With only a few transitions, the rivers and specific places form an uninterrupted line in a soothing, shimmering atmosphere. These lines, both geographical and poetic, function to connect time and space, resembling a line on the mental map formed by Yuan and shared by his readers. This and the other aforementioned spatial techniques would go on to become the foundation of Yuan's powerful poetic exchange with Bai Juyi.

Bai Juyi also worked to fold broad geographic imagination into succinct poetic lines. Although never known as a mapmaker like Yuan Zhen, he was among the most experienced travelers of his time and a keen observer of the cultural geography along the way. Having traveled around southern China with his father in his teens, Bai was again on the road after a brief period of stable service in Chang'an. His journeys covered present-day Jiangxi, Sichuan, Zhejiang, and Jiangsu before leading back to Chang'an and Luoyang in his later years. As he crossed the imperial south, Bai developed his own ways of engaging geography to achieve unique poetic effects. Through them, he reorganized his geographical experiences to fashion his self-image as a traveling poet in a turbulent world.

A vivid example can be found in "Zhuchen Village" ("Zhuchen cun" 朱陳村, c. 808), an autobiographical poem that Bai composed during a trip to retrieve his maternal grandmother's remains.[11] The poem draws a sharp comparison between the idealized peaceful village community and his own turbulent traveling life. The following is the portion of the poem lamenting the author's unsettled life:

> I recall the time I first left my home;
> From then until now—fifteen springs!
> My lonely boat has thrice sailed to Chu;
> Four times through Qin my lean horse has passed.
> I have walked all day with hunger on my face;
> I have lain at night with a soul that could not rest.
> East and west I have wandered without pause,
> Hither and thither like a cloud astray in the sky.
> In the social disorder I lost my old home.
> My family members are mostly scattered.
> North of the River, South of the River,
> In both lands are the friends of my lifetime.

> 憶昨旅遊初
> 迨今十五春
> 孤舟三適楚
> 羸馬四經秦
> 晝行有飢色

夜寢無安魂
|東西不暫住
來往若浮雲
離亂失故鄉
骨肉多散分
江南與江北
各有平生親[12]

This is a highly condensed autobiographical account. Numbers are featured in neatly paired couplets: the fifteen years he spent on the road, the three times he visited Chu, and the five times he passed through Qin; the east and west directions are first paralleled with "hither and thither" and then counterpointed with lands "North of the River and South of the River." The passage portrays the author's personal life and numerous regrets essentially in the form of a mental map, displaying several large geographical territories he had repeatedly visited, each adding to the image of a soul set afloat in boundless space.

Bai develops this theme further in "An Occasional Poem Composed in the Luo Area" ("Luozhong ouzuo" 洛中偶作). The Luo area refers to the region centering on Luoyang, the eastern capital of the Tang. In 824, after his term as the prefect of Hangzhou, Bai was called back to Luoyang to serve as a tutor of the imperial heir apparent. Writing in his fifties, Bai continued his geographically organized self-fashioning, while taking pride in his geographic observations of the places he had traveled:

Five years working in the Hanlin Academy,[13]
Four years governing Xunyang.
One year serving as Prefect of Bajun,
Half a year as an assistant minister at the South Palace.
Two years at Zhilun Gallery.
Three years in the hall of a provincial governor.
All told fifteen years,
And my poems number more than a thousand.
My vision rises to include multitudes of images,
Recording the local cultures of all four directions.
Only in the Luo area have I not composed any poems,

And I cannot help but feel regret.
Now I serve as the official in charge of the spring palace
And have started to travel the area.
I wandered along the Yi River
And caught glimpses of Shaoshi Mountain.
When I encounter an object I chant a poem.
When I chant a poem I drink from my cup.
My brush redeems the regret
And my volume of poetry fills the absence.
From time to time I look at myself and laugh
Because my vision is blurry, my temple-hair and beard are grey.
I will be unaware of my coming old age,
and still indulge my wide poetic passions.

五年職翰林
四年涖潯陽
一年巴郡守
半年南宮郎
二年直綸閣
三年刺史堂
凡此十五載
有詩千餘章
境興周萬象
土風備四方
独无洛中作
能不心悢悢
今为青宫長
始来遊此鄉
徘徊伊澗上
睥睨嵩少傍
遇物輒一詠
一詠傾一觴
笔下成釋憾
卷中同補亡
往往顧自哂
眼昏鬚鬢蒼
不知老將至
犹自放詩狂[14]

In the first half of the poem, the places Bai traveled or the official posts he served, the years he spent, and his growth as a poet are organized into a procession of couplets. This passage gives an admirable list of positions scattered in various regions, including Jiangzhou where Xunyang River is located, Sichuan where Bajun is located, Chang'an the capital, and Hangzhou where he was a prefect. The list of official positions and place names is delivered with a sense of vicissitude and adventure, but the two concluding couplets smooth out the rough edges of this turbulent political career. The saving grace of his long, troubled journey across officialdom was his geographic accomplishments: he studied local customs, observed indigenous cultures, learned about geographic features, and was thus confident enough to poeticize the "multitudes of images" (*wanxiang* 萬象) and "indigenous cultures" (*tufeng* 土風) of "all four directions" (*sifang* 四方). As the critic Chen Yunfeng points out, these lines encapsulate the close relationship between Bai's geographical experiences and the subject matter of his poetry.[15]

The poem then turns dramatically from past experience to the prospects of the future. A dutiful official, the author nonetheless makes no political statements here. His ambition is plain and simple: to write as many poems about the geographic features of Luoyang as he has done for his past places of service. He will write about local landscapes like the Yi River (Yijian 伊澗) and Shaoshi Mountain (Songshao 嵩少). He will go around the city and devote a poem to each site he finds noteworthy. He will write exhaustively, as if completing a mini encyclopedia about places and cultures, and only in this way can he be relieved of his regrets. Ultimately, as the conclusion proclaims, poetry writing as such will embody the meaning of his life: "I will be unaware of my coming old age, / and still indulge my wide poetic passions." As a whole, the poem continues to explore the topic of "Zhuchen Village," exemplifying Bai's multiyear quest for order, meaning, and modes of transcendence in a disorienting age, as well as his conscious and continuous determination to poeticize that quest. He eventually finds an anchor for his relentless quest in writing about geography and culture: another apt reflection of the heightened geographical awareness in the mid-Tang era.

GEOGRAPHICAL SYNCHRONY IN POETIC SPACE

The vibrations of spatial-geographical elements in the poetic realm gave Yuan Zhen and Bai Juyi a unique touch in their individual composition. It was in their corresponding writing, however, that such elements blended and grew into a full-blown literary phenomenon. These elements enabled the poets to carve out a new literary space, in which they cultivated their legendary friendship in dramatic scenes of poetic synchrony.

The poems that materialize this geographical synchrony are "Dream in Liangzhou" by Yuan and "Drunk and Missing Yuan the Ninth with Li the Eleventh" by Bai. They have been circulated as a set of corresponding poems, but have been most dramatically framed in a story written by Bai Xingjian 白行簡 (776–826), Bai Juyi's brother. To fully appreciate the sense of drama the two poems aroused among their contemporary readers, I would like to introduce them first within Bai Xingjian's narrative.

Bai Xingjian's story was grouped together with two other mystery tales under the collective title "Record of Three Dreams" ("Sanmeng ji" 三夢記).[16] The story tells how Yuan Zhen dreamed about Bai Juyi's excursion in Chang'an with surprising precision while he was on his way to Sichuan; Bai Juyi composed a poem on exactly the same date about Yuan Zhen's travel:

> In the fourth year of the Yuanhe Reign period Yuan Weizhi [Yuan Zhen] of Henan, who was serving as Examining Censor, was sent on a mission to Sichuan. Some ten days later my elder brother Letian, Li Biaozhi of Longxi [in modern Gansu Province], and I were on an outing together to the Qujiang. We got to the compound of the Maternal Kindness Temple, visited all of the quarters, and lingered there for a long time. After it grew dark, we went together to Baiozhi's house in the Xiuxing Ward where we ordered some wine, toasted each other and had an extremely enjoyable time

drinking together. My brother put down his cup for a while and said, "Weizhi must be at Liaozhou by now." He asked me to inscribe his poem on the wall of the room. It reads as follows:

> Spring comes, no way to soothe the melancholy of spring.
> Drunken, I break some flowering boughs to count
> the wine cups with,
> And suddenly recall my old friend traveling afar.
> Counting the days and miles he has travelled, I suppose today
> he must have reached Liangzhou.

It was then the twenty-first of the month. About a fortnight later an envoy from Liangzhou arrived and we received a letter from Weizhi. At the end of the letter there was a postscript with a poem entitled "Poem Recording a Dream." It reads:

> In my dream, you brothers were on the headstream
> of the Qujiang,
> And went on to visit the yard of the Maternal Kindness Temple.
> My subordinate officials ordered the servants to have
> the horses ready.
> When I awoke I was surprised to find myself in old Liangzhou.

The day and month of this poem were exactly the same as those of the poem written on the wall after the excursion to the temple. This is what is called a dream in which one person is doing something and another person dreams about it.[17]

元和四年，河南元微之為監察禦史，奉使劍外。去踰旬，予與仲兄樂天，隴西李杓直同遊曲江。詣慈恩佛舍，遍歷僧院，淹留移時。日已晚，同詣杓直修行里第，命酒對酬，甚歡暢。兄停杯久之，曰：“微之當達梁矣。”命題一篇於屋壁。其詞曰:

“春來無計破春愁，醉折花枝作酒籌。忽憶故人天際去，計程今日到梁州。”

實二十一日也。

> 十許日，會梁州使適至，獲微之書一函，後寄《記夢詩》一篇，其詞曰：
>
> "夢君兄弟曲江頭，也入慈恩院里遊。屬吏喚人排馬去，覺來身在古梁州。"
>
> 日月與遊寺題詩日月率同，蓋所謂此有所為而彼夢之者矣。[18]

In Bai Xingjian's "Record of Three Dreams," the two other stories are written in the style of records of anomalies (*zhiguai* 志怪). They tell how multiple protagonists had interrelated dreams without prior coordination, but neither involves the author as a witness.[19] For readers of the mid-Tang, the grouping of the Yuan-Bai telepathy with the other two stories would have conveyed a heightened air of mystery.

The two poems that were the cause of much wonderment are embedded nicely in the short text. The versions of them in Yuan and Bai's collected works have a few minor textual variations. I present them in parallel in figure 2 to facilitate the analysis that follows.

Figure 2. Comparison of Poems by Yuan and Bai.

"Dream in Liangzhou"
by Yuan Zhen

In my dream, you brothers were on the headstream of the Qujiang,

And went on to visit the Maternal Kindness Temple.

The manager of the posthouse ordered the servants to lead the horses.

When I awoke I was surprised to find myself in old Liangzhou.

夢君同繞曲江頭
也向慈恩院院游
亭吏呼人排去馬
忽驚身在古梁州

"Drunk and Missing Yuan the Ninth with Li the Eleventh"
by Bai Juyi

When flowers bloom, we all get drunk to soothe the melancholy of spring.
Drunk, I break some flowering boughs to count the wine cups with,
And suddenly recall my old friend traveling afar.
Counting the days and miles he has travelled, I suppose today he must have reached Liangzhou.

花時同醉破春愁
醉折花枝作酒籌
忽憶故人天際去
計程今日到梁州

Sources. Yuan, *Yuan Zhen ji biannian jianjiao: shige juan*, 145; and Bai, *Bai Juyi shiji jiaozhu*, 14.1067.

Even outside the context of Bai Xingjian's story, the synchronistic nature of these two poems would have appeared remarkable to the mid-Tang reading public. In Yuan Zhen's work, he enters a dream where the Bai brothers are wandering the scenic Qujiang waterfront of Chang'an. But as he is woken up by the manager of the posthouse ordering the servants to prepare the horses, he finds himself stuck half a world away in Liangzhou, an ancient province whose name appeared in the classic "Tribute of Yu." In Bai's corresponding piece, what appeared to be Yuan's nocturnal fantasy transpires as reality in Bai's life on that very day. He tours the lakeside with his brother, drinking to disperse the melancholy of the lonely spring. The image of drinking builds up a sentimental atmosphere while giving the impression of spontaneity. The poet then shifts his attention to his friend Yuan Zhen, who is on the road to Sichuan, and the result of the switch is a matching of geographic location and poetic rhyme: the ancient Liang Province shows up again, linking the two companions ever more closely in the literary imagination.

What enables the interpersonal resonance over the remote distance between the capital and the periphery are a set of temporal-spatial techniques, knowledge, and tricks that are reflected in Yuan Zhen, Bai Juyi and Bai Xingjian's respective writings. The first is the two poets' mutual tracing of each other in their own mental maps. Both poems were in essence a series of motions through recollected space: in his poeticized dream, Yuan Zhen visualizes the Bai brothers as following a travel route in the Qujiang area, walking from the headstream where the literary elites would gather at the Maternal Kindness Temple. He then plays with distances, coupling the Bai brothers' short excursion to a bustling site of metropolitan pleasure with the faraway, archaic province where he temporarily resides. A host of motion verbs riddle the mental image of Qujiang, but all movement comes to a sudden halt with the static word "在," or "to be at" in the last line. The speaker lies still, and his presence in the poem is underlined by his absence from the pleasant scene in Qujiang.

Bai Juyi's poem shares with Yuan's poem the same vision of the relationship between Qujiang and a midway posthouse projected by a geographer's sharp mind. But here the poem carefully veils it with an ordinary drinking scene. The speaker sits back under blossoming flowers, tipsy and melancholic, busying himself with a series of counting activities. He breaks flower branches to count the cups that he has emptied in pensive sadness, before realizing that it is his soulmate's absence that has deepened his sorrow. So he sets the flower branches aside, turning to count the miles Yuan must have covered on the road to distant Sichuan and calculating the midway points he would have reached each day.

Both Yuan and Bai thus fold movements through large-scale spaces into a poeticized mental map, hinting at their affectionate interaction through the contrastive interplay of routes, distances, and locations both ancient and contemporary.

The second set of temporal-spatial factors that contributes to the sense of synchronic resonance in the two poems is to be found in Yuan and Bai's subtle understanding of specific places. Their geographical acuteness offers a plausible empirical explanation for the ostensibly supernatural coincidence. In Yuan's poem, he demonstrates a careful selection of geographic references. In his poetic dream, he chooses to set the imaginary travel scene in Qujiang. The choice is likely deliberate, for Qujiang held special significance for the pair and for the larger Tang literary community. To borrow the sociologist Michael Farrell's term, Qujiang was a "magnet place," or a place where highly ambitious members of a discipline create their collaborative circle.[20] The area stretched around a manmade lake called Qujiang Pond, and was decorated with some of the best gardens in the capital. Several Buddhist temples, including the "Maternal Kindness Temple" Yuan invoked in the poem, lent the lakeside further transcendental appeal.[21] In the mid-Tang, the area grew into a center of the literati community through its function as the site of celebratory feasts and gatherings after the civil service examination.[22] The examination drew political aspirants from all over the empire, and Qujiang attracted the most talented among them. There, the newly selected candidates for official posts would seek public recognition and network with their peers. The vogue quickly spread beyond the examinee circles into the larger literary community.[23]

Yuan Zhen and Bai Juyi evoked the Qujiang area many times in their poems written while they were still both in Chang'an. They used to stroll together in the scenic site, polishing their poetry and debating politics. The area served as a symbol of their friendship and for the larger literary culture in which they took part. Yuan's earliest extant poem "One Hundred Rhymes Written on Behalf of the Old Man of Qujiang" ("Dai

Qujiang laoren baiyun" 代曲江老人百韻) uses the Qujiang area as the setting for a long narrative about the Tang empire's history since the Rebellion. In another poem with a long title, "On February Two of the Second Year of the Yongzhen Reign, the Emperor Went to the Red Phoenix Tower to Announce the Amnesty to the World. I Went on a Leisurely Walk by the Qujiang Pond with Li Gongchui [Li Shen] and Yu Shunzi [Yu Jingxiu] and Thus Missed the Grand Event" ("Yongzhen ernian zhengyue erri Shang yu Danfeng lou she tianxia yu yu Li Gongchui Yu Shunzhi xianxing Qujiang buji shengguan" 永貞二年正月二日上御丹鳳樓赦天下予與李公垂庾順之閒行曲江不及盛觀), Yuan summoned the image of himself touring the symbolic site of Qujiang with his friends as an expression of their collective resilience in the face of precarious court politics.[24] Yuan and Bai continued to feature the site in their poems after their encounter in Chang'an, with Bai leaving a staggering body of sixty Qujiang-related poems.[25] Therefore, when Yuan called upon Qujiang in his poem, he was salvaging colorful memories of the past in a displaced present.[26] The mere mention of the place would have triggered emotional resonance in his poetic partner, inviting a poetic response. We may read it as the hallmark of a designed poetic exchange, even though in this particular case Bai's response somehow preceded the invitation. The reference would have also appealed to Yuan's broader audiences, as they all partook in the culture of Qujiang.

If Yuan Zhen's invocation of Qujiang was a deliberate cultural act aimed at creating interpersonal resonance, Bai Juyi's precise guess about Yuan's whereabouts on that particular day came from his broad knowledge of Tang geography and the culture of travel. As I have framed it in the close reading earlier, both poets had in their mind a mental map of their soulmate's movements across time and space. Whereas Yuan's map was more sentimental, Bai's geographic thinking was empirical, presenting the image of an experienced geographer concerned about his friend. Studies of Tang transportation systems show that the state had fairly reliable control over official travel routes across the south.[27] For example, imperial geographical documents by officials such as Li Jifu had a thorough description of all the main transportation hubs and routes, in addition to the distance between each major location and those in its sixteen directions.[28] Tang map-guides also contained detailed information about the posthouses, which provided accommodation and transportation services to traveling government officials. For instance, in *Map-Guide of the Shazhou Military Commandery*, the category of posthouse (*yi* 驛) has the largest number of entries.[29] These sources of information enabled the Tang state to meticulously regulate official travel, stipulating the distance one needed to cover in each single day.[30] In other words, when Bai Juyi calculated the days of Yuan Zhen's travel, he must have had access to information about the route that Yuan took.

The third factor that finalizes the synchrony in Yuan and Bai's poems is Bai Xingjian's story. Although not involving any more geographical understanding than what the two poets showed in their poems, Bai Xingjian's story provided additional context that was crucial to shaping contemporary readers' reading experience of the two poems. Within the two poems themselves, there is in fact little evidence to show that the Bai brothers were at Qujiang when Bai Juyi wrote his poem; nor do we know for certain that Bai and Yuan's poems were written on the same day. We know all this only from Bai Xingjian's story. Moreover, by grouping this story with the other two stories written in the style of records of anomalies, Bai Xingjian invited readers to associate the two poems with any unknown, magical power of the likes that are commonly seen in records of anomalies. In other words, the effect of synchronization that Yuan and Bai created in the two poems is in part contingent on Bai Xingjian's narrative reframing of them. The narrative supplements the formal restriction of the short poems and fills in the informational gaps between the lines. It reframes Yuan and Bai's poetic compositions into a contemporary legend, and thus helped celebrate Yuan and Bai's creative partnership.

As a result, the two places, one in the imperial center and one on the way to the periphery, are put into interaction in both poems. The poets are synchronized at different locations and are united in the textual space. Contemporary readers would have been impressed by the interpersonal resonance and temporal synchrony that came through from poetic text and its narrative context in Bai Xingjian's story; at the same time, they would have also been impressed by the two friends' concern about each other, articulated through their respective envisioning of each other in their mental maps. In this sense, we might say that there are two synchronies involved in Yuan and Bai's two exchange poems and Bai Xingjian's story: the marvelous simultaneity that belongs to the realm of coincidences, and the synergy between poetic creativity and geographic prowess. These worked in concert to conjure up the spatial imaginary of the two poets longing for each other in spite of the long and changing distances that kept them apart.

Comparative Geography and the Diptych Form

We have seen in the previous section how Yuan Zhen and Bai Juyi managed to stay connected in their long-distance corresponding poetry by tracing each other in their mental maps when they were separated. Retrospectively, their corresponding writing during Yuan's official trip to Sichuan might be seen as a prelude to the more sophisticated geographically informed poems the two exchanged during a longer separation and under harsher political circumstances. This section is focused on one set of exchange poems they wrote during the mid-810s, when Yuan and Bai were separated for the majority of the time due to their banishment to the south. The poems are Bai's "Song of the Southeast" and Yuan's "In Response to Song of the Southeast." In Yuan's corresponding poem, he invents a unique method of engaging with geography in poetry through the juxtaposition and comparison of their respective locations in what I call the diptych form. In this form, two poems written in "long-regulated verse" (*pailü* 排律) correspond to each other through the method of "matching rhymes" (*ciyun* 次韻) to build progressive couplets that unfold, as though on two scroll paintings, the two poets' corresponding spatial representations. Assimilating a thoughtful selection of comparable geographical details into this interlocked poetic form, Yuan's exchange poem integrated the two poems and the two places, therefore firmly maintaining the connections between the separated friends and subtly altering the spatial perception of their readers.

Between 810 and 815, Yuan Zhen was demoted to Jiangling in Sichuan, leaving Bai Juyi alone in the capital. In early 815, Yuan was called back by the court to Chang'an, only to be sent to Tongzhou in Sichuan again several months later. In the same year, Bai was demoted as well. He embarked on a long journey to Jiangzhou, which was more than two thousand *li* away from the more backward Tongzhou. Because there was no direct governmental postal service between the two places, they could only rely on their traveling friends to serve as occasional messengers. To make matters even worse, their contact was cut off from October 815 to April 817, when Yuan was stricken by illness in Tongzhou and sent to a different town for medical treatment. Bai continued to write poems and letters addressed Yuan, though he did not know whether Yuan was still alive. As he laments in a letter to Yuan:

> Weizhi, Weizhi, it is three years since I saw your face and almost two years since I had a letter from you. Is man's life so long that he can afford such partings? Much less should hearts joined by glue be set in bodies remote as Hu and Yue [the extreme north and south of China]. In promotion we could not be together; and in failure we cannot forget each other. Snatched and wrenched apart, separately each of us grows grey. O, Weizhi, what is to be done? But this is the work of heaven and there is no use in speaking of it.[31]
>
> 微之微之！不見足下面已三年矣，不得足下書欲二年矣。人生幾何？離闊如此。況以膠漆之心，置於胡越之身，進不得相合，退不能相忘。牽攣乖隔，各欲白首。微之微之，如何如何？天實為之，謂之奈何？[32]

Among the work Bai sent to Yuan during this period was his long poem of one hundred couplets, "Song of the Southeast." The poem gives an all-inclusive depiction of Bai's past and recent experiences, including twenty couplets on his living environment in Jiangzhou. In 817, Yuan returned to Tongzhou and resumed contact with Bai, but most poems Bai had sent him had been lost. Bai therefore transcribed all of them, "Song of the Southeast" being the longest, and sent them to Yuan again.[33] Yuan soon started to reply with corresponding poems. In 818, he responded to this long poem with his "In Response to Song of the Southeast" in matching rhyme words. Yuan's poem contains a substantial section of an interactive depiction of Tongzhou and Jiangzhou, the two places where the poets were located, as a response to Bai's depiction of Jiangzhou in his original poem. In the following, we will first examine how Yuan established a comparative geography of resemblance in his corresponding poem, and then look into how the effect of geographic writing is dramatically reinforced by the diptych form.

In Bai Juyi's original poem, he opens with a succinct summary of his trip from the capital to Jiangzhou: "Going south I passed the three Chus. / Heading to the east I passed the Five Lakes" (*Nanqu jing san Chu, donglai guo Wuhu* 南去經三楚, 東來過五湖).[34] In Yuan Zhen's corresponding poem, he starts with a couplet that introduces his own circumstances in the first line, and refers to Bai's trip in the second: "I was sick and close to death, / and you have already crossed the lake" (*Wo bing fang yin Yue, Junxing yi guo Hu* 我病方吟越，君行已過湖).[35] In this way, Yuan brings two people at two locations together into one couplet with its parallelism. Yuan's first couplet thus anticipates the kind of correspondence that will later reoccur in the section of the poem about Jiangzhou and Tongzhou. In Bai's poem, he then occupies several lines with his appreciation of the landscape on his trip to Jiangzhou. In Yuan's correspondence, he uses the same number of lines to delineate a lonely self-portrait. Then, as Bai starts to depict the local geography of Jiangzhou, Yuan inserts a note into his poem: "After this point, in each couplet, the first half deals with the local customs in Ba and Shu, while the second deals with the local produce of the Yangtze Valley" (*Cihou meilian zhinei, banshu Bashu tufeng, banshu Jiangxiang wuchan* 此後每聯之內，半述巴蜀土風，半述江鄉物產).[36] Ba and Shu generally refer to the area of modern Sichuan, and Ba specifically refers to the southeastern part of Sichuan, including Tongzhou.

With that, Yuan Zhen begins a lengthy engagement with the geography of Jiangzhou and Tongzhou. In the corresponding part in Bai Juyi's original work, he first states that the local land of Jiangzhou is different from the northern heartland: "I gradually felt the differences in the local lands / And deeply recognized how the produce differs" (*Jianjue xiangyuan yi, shenzhi tuchan shu* 漸覺鄉原異，深知土產殊).[37] Yuan answers in his poem, "I feel the pain in vain as my bones and muscles hurt, / at the same time, you exclaim that the produce and climate are different" (*Zutong jinhai can, pangjie wuhou shu* 坐痛筋骸憯, 旁嗟物候殊).[38] We will soon know, however, that Bai's highlighting of the geographical difference between Jiangzhou and the northern heartland will eventually be replaced by Yuan's conclusion of similarities between Jiangzhou and Tongzhou, the two southern peripherals where they were located.

Bai continues on to describe the sound of Jiangzhou's local people:

> The aboriginals' language babbles and burbles.
> With barbaric expressions, they laugh with excitement.
>
> 夷音語嘲哳
> 蠻態笑睢盱[39]

In response, Yuan's couplet compares Tongzhou's buzzing insects in the humid environment and Jiangzhou's water monster that braves the waves:

> Steamy rain stirs the noisy insects.
> The surging waves inspire the water monsters' excitement.
>
> 雨蒸蟲沸渭

浪涌怪睢盱[40]

As we can see, while the first line of Yuan Zhen's couplet is meant to engage with Jiangzhou, it is not in any sense a rephrasing of Bai Juyi's couplet. Because Yuan has never been to Jiangzhou himself, his knowledge of Jiangzhou may have come from his own reading, perhaps from Bai's introductions in other pieces of writing. We know from the earlier section of the chapter that Yuan was an experienced reader of geographical works. In any case, what is revealing here is that Yuan did not choose geographical details randomly, but he endeavored to find comparable details for both places so as to imply a similarity. He first introduces the rain and the insects in Jiangzhou, and then pairs that with a picture of Tongzhou that also features abundant water.

Here is another example of this comparative geography. In Bai's poem, he depicts the fauna in Jiangzhou:

> Singing in the mountains, the ape cries alone.
> Crying in the wildness, the birds call in response.

> 山歌猿獨叫
> 野哭鳥相呼[41]

Yuan responds with the linguistic landscape in Tongzhou:

> The aboriginals' sound resembles laughter.
> The barbarian language communicates like riddles.

> 夷音啼似笑
> 蠻語謎相呼[42]

This couplet is apparently a reworking of Bai's earlier couplet "the aboriginals' language babbles and burbles. / With barbaric expressions, they laugh with excitement." And yet Yuan's couplet is meant to depict both places, with the first line attributed to Jiangzhou and the second to Tongzhou. When read in this new framework, both lines highlight the linguistic barrier the two poets experienced in both places. The resemblance being addressed here is not precisely a geographical similarity, as Yuan is not indicating that the dialects of the two places are similar to each other in the linguistic sense. Rather, what is being compared and associated are the two poets' geographical experiences. It is their alienation from the local linguistic environment that links them together.

In this way, in the two corresponding sections on local geography, both Bai Juyi and Yuan Zhen represent local realities in the south, describing various images of landscapes, routes, historical sites, local customs, indigenous peoples, exotic fauna and flora, and so on. Overall, in the matching poem, Yuan splits twenty-two couplets into two separate but interconnected planes: the first line represents the realities in Tongzhou, the second those of Jiangzhou. The special effect is that the couplets fold in parallel descriptions of the two places, and then as a whole match Bai's descriptions of one place. Putting their poems together, the reader sees an unfolding, fractal world, in which two peripheries of the empire are connected by a series of contrasting and comparable points.

Toward the end of this section of interactive literary geography, Yuan goes beyond the details and concludes with some general similarities between the two places on a larger scale:

> The wind of Chu is as light as that of Shu.
> The land of Ba is as humid as that of Wu.
> The air is murky, so stars are hardly seen.
> The land of the prefecture slopes, so the sun drops easily.
> It is cloudy throughout the night.
> Not yet the *you* hour, the sun sets.[43]

> 楚風輕似蜀
> 巴地濕如吳
> 氣濁星難見
> 州斜日易晡
> 通宵但雲霧
> 未酉即桑榆[44]

In this conclusion, the "wind" connotes the "folk culture" that joined Chu (which, in cultural terms, included Jiangzhou) with Shu. The humid environment of Ba pushes Yuan Zhen further afield, stretching his comparative geographic scope to include Wu in the lower Yangtze area. Yuan thereby reframes Jiangzhou and Tongzhou in the larger and more dynamic scope of southern China as a whole, in which the two regions where the poets resided resembled each other in customs and natural environment. In the following lines, Yuan continues to parallel the foggy night in one place with the early dusk in another, indicating the shortage of sunshine in both. Thus, while the geographical sections in the two poems start with Bai exclaiming over the difference between the northern heartland and the Jiangzhou area, they end by turning enumerations of the similarities of the two places into a larger geographical picture of the southern land, encompassing both places.

While Yuan's careful choice of geographical details of both places made his comparative geography possible in the first place, what decisively enhanced the effect of geographical integration between the two places was his employment of the unique diptych form. By diptych form, I mean long-regulated verse with matching rhymes. This form involved both experiment and competition, and distinguished Bai and Yuan's high craftsmanship in versification. When applied to geographical topics related to the poets' experiences, the diptych form became a powerful tool of spatial organization and connection, engaging the geographically separated and otherwise disconnected poets within an intimate and interactive poetic space.

Long-regulated verse is a form of "modern-style poetry" (*jinti shi* 近體詩) established during the early Tang. Its formal characteristics can be described as follows: regulated pentasyllabic (occasionally septasyllabic) couplets of more than ten lines; strict syntactical parallelism in all the couplets with the exception of the opening and concluding ones; all even lines end-rhyme, without repetition of end-rhyme words. The more couplets the poem accrues, the more difficult it is to create progressive parallelism with rhymes; in addition, the longer the poem runs, the more the writing tends to be open-ended and requires poetic improvisation at every turn. What emerges is that long-regulated verse turns out to be an exercise in arranging rhymes, a reworking of traditional themes and diction, and a sustained expansion of the poet's thematic scope.[45] Du Fu was the first poet to write "long-regulated verse of one hundred rhymes" (*baiyun pailü* 百韻排律), consisting of two hundred lines and one thousand characters, which came to stand as the highest standard of poetic craftsmanship in long-regulated verse. Following Du, Bai Juyi and Yuan Zhen were the only mid-Tang poets who managed to write such long poems in this form. Unlike Du Fu, who worked as an individual poet, Bai and Yuan wrote correspondingly, which makes their long-regulated verse all the more impressive.

To make their poetic experiment even more challenging, Bai and Yuan added matching rhymes to the composition. "Matching rhymes" mandated that one poet compose a piece, and then the other poet use the same rhyme words in the same order to compose a matching poem. This intricate compositional method suggests an intimate relationship between the poets, at once collaborative and competitive. Yuan and Bai first attempted to use this method in exchange poems in 810, and then continued to develop techniques for it in various lengths in the Tongzhou-Jiangzhou period. According to the literary historian Chen Caizhi, there are 58 extant groups of exchange poems composed by Yuan and Bai from 815 to 819, and 36 of them use matching rhymes.[46]

Here is a quick example to show how these rhymes work in an ordinary pentasyllabic poem that does not employ high formalism. Bai's poetic sequence "Three Poems Sent to Weizhi" ("Ji Weizhi sanshou" 寄微之三首), written in 815, starts with the following lines:

> From Jiangzhou to Tongzhou
> From the edge of the sky to the earth's end.
> The mountains are ten thousand *zhang* high,
> The water is one thousand *li* wide.

Clouds and mist further separate us
A distance even birds cannot cross.

江州望通州
天涯與地末
有山萬丈高
有水千里闊
間之以雲霧
飛鳥不可越[47]

Yuan's matching poem, composed in 818, starts with:

In the past we lived in the center of the capital
Now we are at the edges of Wu and Chu.
The roads through thousands of mountains are dangerous,
And for our correspondence ten thousand *li* is too far.
As far apart as Orion and Antares,
On earth the distance is like that between Hu and Yue.

昔在京城心
今在吳楚末
千山道路險
萬里音塵闊
天上參與商
地上胡與越[48]

For Yuan Zhen, the last character of each couplet is already fixed. He has to move backwards to write the other four characters, both arranging them to counterpoint the meaning of each of Bai's lines and progressing forward with a coherent meaning of his own. If he fails to create a satisfactory matching line or if he desires more breadth, he can explore the fixed character's other meanings. For instance, the word *yue* in Bai's poem is a verb meaning "to cross," and in Yuan's poem it is a place name that stands for the far south.

When "long-regulated verse" is joined with "matching rhymes," the challenge multiplies. In this case, the two poems hinge upon each other while displaying extended development, as if a diptych is being unfolded.

When the verse reaches a hundred rhymes, a diptych form is a daunting task for any pair of poets to accomplish. Understandably, Yuan and Bai left only two pairs of diptych poems with a hundred rhymes. The first pair was written in 810. Bai's "A Poem of One Hundred Rhymes Sent to Weizhi in Place of a Letter" ("Daishu shi yibai yun ji Weizhi" 代書詩一百韻寄微之) was composed in Chang'an, while Yuan's "In Response to the Hanlin Scholar Bai's Poem of One Hundred Rhymes Sent to Weizhi in Place of a Letter" ("Chou Hanlin Bai xueshi daishu yibai yun" 酬翰林白學士代書一百韻) was composed in Jiangling. The second pair is "Song of the Southeast" by Bai and "In Response to Song of the Southeast" by Yuan. Limited by the formal challenge, poems in diptych form can hardly evolve around a singular, narrow theme; rather, all four of these poems deal with a range of topics, including the two poets' personal experience, their moral and political sentiments, their appeal to the broader literati community for sympathy and support, and so on. Therefore, when Yuan Zhen devoted forty-some lines to the fixed topic of Jiangzhou and Tongzhou's geography in diptych form, he set the bar of poetic challenge unprecedentedly high.

The result of this challenge is an intimate and organic integration of the two places. The juxtaposed quotes in figure 3, starting with the lines we analyzed earlier, can give us a sense of this compelling effect.

Figure 3. Juxtaposed Quotes by Yuan Zhen.

"Song of the Southeast"

The aboriginals' language babbles and burbles.

Misty villages merged with clusters of boats.

Clerks levied fishermen's taxes.

Peasants paid taxes for their slash-and-burn farms.

The *hai* days signaled harvests of shrimp and crabs.

The *yin* years welcome to the land plenty of tigers and cheetahs.

夷音語嘲哳
蠻態笑睢盱
水市通闤闠
煙村混舳艫
吏征漁戶稅
人納火田租
亥日饒蝦蟹
寅年足虎貙

"In Response to Song of the Southeast"

Steamy rain stirs the noisy insects.

The surging waves inspired the water monsters' excitement.

Mosquitoes and blackflies flew around boat-ropes.

Tumbleweed and hemp covered clusters of boats.

Straw thatches the low roofs.

My meager salary just matches a fisherman's tax.

On the muddy riverbank clamorous people collect clams.

In the desert men fight with savage cheetahs.

雨蒸蟲沸渭
浪涌怪睢盱
索綆飄蚊蚋
蓬麻甃舳艫
短檐苫稻草
微俸封漁租
泥浦喧撈蛤
荒郊險鬥貙

In this quote from Bai's poem about Jiangzhou, following a depiction of the unintelligible dialect, Bai goes on to observe the human society established within the system of waterways. He then switches his attention as a newly arrived official to the ways the government rules the local people through taxation. Finally, he introduces the folkways relating to local aquaculture and hunting according to the calendar.

In Yuan Zhen's corresponding poem, he uses two couplets to compare the similarly water-bound environments and customs of Jiangzhou and Tongzhou. Yuan then follows with a juxtaposition of the material features of the living environment in Jiangzhou with his own frugal life as a local official in Tongzhou. In the final couplets, he compares and associates two scenes of men striving and surviving in the wild in both locations.

The diptych form connects both the two poems and the two places on multiple levels. We have discussed that within Yuan's couplets, the depictions of both places correspond with each other. Now, between the two poems, the same rhyme words at the end of every other line hinge the two poems together, further combining the two places and highlighting their similarities in the form itself. Moreover, the sheer length of the geographical parts in these poems, typical of long-regulated verse, gives the readers a strong sense of progression. While Bai unfolds the geographical features in Jiangzhou successively in his poem, Yuan provides the reader with a long list of comparisons before drawing the conclusion about similarity. This creates a reading experience that resembles a journey of curious geographical exploration and discovery, and makes Yuan's conclusion all the more convincing. Whether reading Yuan's poem alone or reading the two poems next to each other, one cannot miss the clear message that the two otherwise irrelevant places are connected in multiple ways, are corresponding with each other, and are integrated as a whole in Yuan and Bai's poetic space.

This interactive geography kept the two poets connected despite the obstacles and helped to consolidate their friendship. At first, the daunting distance between Jiangzhou and Tongzhou separated the two friends physically and psychologically. As Bai writes in the poem "Three Poems Sent to Weizhi" mentioned earlier, when the friends first arrived in their respective peripheral posts, they felt that they had gone from "the edge of the sky to the earth's ends." Between Jiangzhou and Tongzhou were high mountains, wide rivers, clouds and mist that "even birds cannot cross." But two years after their separation and disconnection, Yuan Zhen was able to establish an intimate relationship between the two places in his long exchange poem through comparative geography. The interaction and integration of the two areas in Yuan's poetic space form a sharp contrast with the unconquerable geographical distance that Bai lamented earlier. Moreover, the diptych form also embodies Yuan's desire to maintain his connection with Bai over a long distance. In the preface to his matching poem, Yuan explains:

> On April 13, I wrote the poem in two volumes and continued to repeat the original rhyme words in their original order. I also do not know when I will be able to see Bai, and so I will probably rely on other people to bring them to him. In Tongzhou there is no one with whom I can talk about poetry, and only my wife Shu understood me.
>
> 四月十三日，予手寫爲上下卷，仍依次重用本韻，亦不知何時得見樂天，因人或寄去。通之人莫可與言詩者，唯妻淑在旁知狀。[49]

Here, Yuan's corresponding writing with matching rhymes becomes a compensation for the pain he feels over not being able to see Bai in person, as well as over the lack of a satisfying intellectual environment in Tongzhou. In other words, his formal engagement in poetry composition becomes an important emotional articulation.

To sum up, the comparative geography in the diptych form enabled Yuan to replace isolation with communication and staged a repeated reunion with Bai in the textual space. For their readers, the two poems presented the poets as an inseparable whole, through their corresponding writing and their shared geographical vision of the imperial south.

This approach of comparative geography was occasionally also employed in contemporary geographical works. A comparison between a geographical case and Yuan-Bai's poetry will further elucidate the interpersonal nature of the latter. In the *Map-Guide of Shazhou*, the compiler compares the flora of Shazhou in the northwest region with that in the remote eastern China:

> The above-mentioned Shazhou belonged to the territory of ancient Guazhou. Its land consists of plains, and the soil is mostly sandy and alkaline. The locals are farmers. The local flora is similar to that in eastern China, with the exception of idesia, the parasol tree, the catalpa tree, the lacquer tree, the Chinese juniper, and the cypress.
>
> 右，沙州者，古瓜州地。其地平川，多沙鹵。人以耕稼為業。草木略與東華夏同，其木，無椅、桐、梓、漆、栝、柏。[50]

This passage, like Yuan and Bai's poems, compares the geography of two places distant from each other. However, the purpose of comparison is to introduce the local geography in the western frontier of the empire without having to list all the local plants and trees. For this purpose, the compiler compares a place in the empire's northwestern frontier with a place that the general readers of the map-guides, possibly the local officials newly appointed by the central court from outside the local region, would be more familiar with. In contrast, the comparative geography in Yuan Zhen's poem has little geographical foundation except for the fact that the two places both belonged to southern China. Comparing Jiangzhou to Tongzhou would hardly make it easier for the readers to grasp the geographical features of either place, because most potential readers of the poems, especially those in the capital, would be unfamiliar with both. Moreover, sometimes the comparison in the poems is not between geographical features of the two places, but rather between the poets' experiences with local geography, such as their common alienation from the local linguistic milieu. The essence of Yuan and Bai's geographical writing in exchange poetry, then, is to establish a mental route between two otherwise unrelated places as the embodiment of the personal connection between the poets, as well as to stage their dramatic collective presentation for the broader literati community.

THE ROUTES OF TRAVELING TEXTS AND THE EMPIRE OF POETRY

In the tomb inscription Bai Juyi wrote for Yuan Zhen following his death in 831, Bai affectionately recalled his literary talent and the popularity of his work. In particular, he outlines a geographical range within which Yuan's literary texts were known to have been circulated:

> From the six palaces, two capitals, and regions in eight directions to the southern *man* barbarians' land and the eastern *yi* barbarians' kingdom, [Yuan Zhen's poems] were copied and circulated everywhere. When a new piece or new line of his came out, it was off and running without legs, taken away faster than pearls and jade.
>
> 自六宮兩都八方至南蠻東夷國，皆寫傳之。每一章一句出，無脛而走，疾於珠玉。[51]

In this demarcation of a spatial scope that contains the circulation of their literary texts, Bai moves from the imperial center to the remote peripheries to beyond the range of the empire. It is as if Bai is drawing a verbal map to illustrate the dissemination of Yuan's literary texts. A map of literary dissemination is not easy because, as Bai puts it, literary texts could take off "running without legs." It is therefore difficult to trace the routes, calculate the speed, and anticipate the destination of a traveling text. While such conditions would frustrate any human traveler, in literary dissemination, the very unpredictability of the travel could thrill readers, boost the author's ambition, and add another layer of charm to the traveling text itself. In this section I will look into how Yuan and Bai worked collaboratively to expand the empire of their literary writing and to dramatize the routes of dissemination within this empire through various literary activities. All these activities culminated in the unprecedented popularity of their work, as is described in Bai's tomb inscription for Yuan.

In June of 815, a demoted Yuan Zhen arrived at his temporary housing in Tongzhou, after spending three months on the road in poor health.

When he arrived at the desolate house at dusk, he found suspicious traces of an animal and was seized by the idea that they belonged to a tiger. Not knowing whether the creature was still lying in wait, he surveyed the house, and suddenly, as if by magic, there appeared on a pillar a poem by Bai, composed fifteen years earlier in 800 to celebrate the life of sensual enjoyment in the capital. Yuan records this unexpected discovery of his friend's old text in the poem "On Seeing Letian's Poem" ("Jian Letian shi" 見樂天詩):

> Arriving in Tongzhou as the sun was sinking to the west,
> The house by the river was empty, with tiger's tracks in the mud.
> Suddenly across the ruined eaves and fractures,
> I saw your poem inscribed on the center of a pillar.

> 通州到日日平西
> 江館無人虎印泥
> 忽向破檐殘漏處
> 見君詩在柱心題[52]

This concise quatrain captures the moment of Yuan's encounter with Bai's poem in an unexpected place and connects itself with Bai's poem. The poem, like other examples of Yuan and Bai's exchange poetry, establishes a personal connection over a long distance. Moreover, it uses swift transitions to contrast Bai's poem with the bleak environment, thus indicating the longstanding popularity of Bai's work. From Yuan's poem, readers do not learn how the fifteen-year-old text had spread from the capital to the deep Sichuan basin, and ended up being copied onto the pillar of a random house. But it is precisely this unknowing that creates the moment of surprise in Yuan's poem and lends a sense of power to Bai's poem.

Yuan Zhen then sent the poem to Bai Juyi, who was still in the capital. Bai retold the story in his own poem with a long title in the form of a preface:

On the day Weizhi arrived in Tongzhou to accept his new office but was not yet settled in, he saw a few lines on a dusty wall. On reading them, he saw it was an old poem of mine, whose last lines read, "On green waters, one bud of red lotus opens; / [against which] the ten thousand flowers and hundred grasses have no beauty," but he didn't know who it was that had inscribed it. Weizhi sighed over it, but feeling that that wasn't enough, he patched together a verse and added my poem and sent them both to me. On examining the poem, [I realized] it was none other than a quatrain I had presented to the Chang'an entertainer A Ruan fifteen years ago, when I passed the exams. As I thought over these bygone affairs that seemed distant, as if in a dream, I was moved by reflecting on the past, and thus now respond with a poem in the long line.

That time from fifteen years ago seems like a dream-journey –
yet once I used these lines of verse to mix with the fashionable.
It happened that I made a funny song to tease A Ruan,
now I know it was passed along and recited as far as Tongzhou.
Back then it made the red sleeved beauty sing –
now it may dispel the blue-robed marshal's sorrow.
But I was sad to hear as well of where it was inscribed:
on top of a broken wall of a rain-soaked river hostel.[53]

微之到通州日授館未安見塵壁間有數行字讀之即僕舊詩其落句云綠水紅蓮一朵開千花百草無顏色然不知題者何人也微之吟歎不足因綴一章兼錄僕詩本同寄省其詩乃是十五年前初及第時贈長安妓人阿軟絕句緬思往事杳若夢中懷舊感今因酬長句

十五年前似夢遊
曾將詩句結風流
偶助笑歌嘲阿軟
可知傳誦到通州
昔教紅袖佳人唱
今遣青衫司馬愁
惆悵又聞題處所

雨淋江館破牆頭[54]

As Anna Shields comments, "...[T]he detailed recounting of the title seems aimed at readers other than Yuan Zhen. Even though Bai claims that the 'bygone affairs seem' 'as distant as in a dream,' the inscribed poem in 800 reclaimed Bai's past popularity—mixing with famous Chang'an performers after he had passed the *jinshi* examination—and documents his continuing fame—which survived, however humbly—'as far as Tongzhou.'"[55] In 800, Yuan and Bai were both rising stars in the cultural scene in Chang'an. Although they had not yet met, the wide circulation of their poems would soon bring them together as competitive collaborators. From a temporal perspective, Bai's poem establishes a link between past and present, expressing Yuan and Bai's shared nostalgia for their successful youthful years. Behind the nostalgic tone, however, we can sense Bai's satisfaction in the uncanny omnipresence of his poetry. Remote Tongzhou where tigers roamed is thus connected, perhaps for the first time, to Chang'an's banquets and music performances. Thinking along this line of spatial connectivity, Yuan's poem may be about a personal matter, but Bai's poem with its preface-length title broadcasts a piece of pleasant news to a much wider public. The distance between Tongzhou and Chang'an can be easily mapped or arduously traveled, but the orbit of Bai's poem's circulation to Tongzhou will always be a mystery. As the poem comes back to Chang'an via Yuan's letter, it completes its return journey.

If Yuan's encounter with Bai's poem copied by some anonymous reader in Tongzhou is a pure coincidence, then at other times, the two poets copied their own poems in unexpected corners of the empire with the anticipation that they would be carried further. In 815, when Yuan was cut off from Bai, he copied one of Bai's poems on the wall of a Buddhist temple in Langzhou in Sichuan, and followed it with a quatrain commemorating this experience, titled "Copying Letian's Poems on the Walls of the Kaiyuan Temple in Langzhou" ("Langzhou Kaiyuan sibi ti Letian shi" 閬州開元寺壁題樂天詩).

I missed you so much that I copied out your poems.
I finished a thousand lines, but whom could I tell?
They are written on the walls of Langzhou's East Temple.
Who can say when I shall see you again?

憶君無計寫君詩
寫盡千行說向誰
題在閬州東寺壁
幾時知是見君時[56]

Two years later, after Bai resumed contact with Yuan Zhen, he read Yuan's poem and responded with two poems. The first, titled "In Response to Weizhi" ("Da Weizhi" 答微之), matches Yuan's copying of Bai's poems with Bai's own copying of Yuan's work on a screen and echoes Yuan's sense of loss as well as his longing for reunion:

You wrote my poems across the temple walls,
I filled a folding screen with your lines.
I do not know where we shall meet again,
Two weeds floating in the sea.

君寫我詩盈寺壁
我題君句滿屏風
與君相遇知何處
兩葉浮萍大海中[57]

The floating weeds symbolize life's transience and serendipity, while the sea denies direction and reunion. Like the weeds, the poets found themselves powerless against the outside forces that decided their life course. To resist this force which had led to separation, they copied each other's poems both in public and in private.

In his second corresponding poem, Bai provides a new possibility, using the same occasion of copying Yuan's poems on the screen:

In winter of the twelfth year [of the Yuanhe Reign], Weizhi is still cast away in Tongzhou, and I have not yet left the Pen River.

Separated by ten thousand *li*, we haven't met for three years. Depressed and longing for each other, we mostly seek to comfort ourselves by chanting poems. Over these years Weizhi has sent me several hundred poems. Although collecting them in a rattan box has preserved them well, it is even better to place them to the left of my seat, which makes me feel as if I can see his thoughts. For this reason, I selected some short and brilliant pieces of his regulated verse, and inscribed one hundred of them on a screen. Reading them I can communicate with his mind as though he were in front of me. Our predecessors mostly acted out of contingency. Yet who knows whether this screen will be circulated by enthusiasts, making my actions a legend of Jiujiang? So I inscribe this quatrain to encourage them.

Missing you, I selected your poems to make this screen.
I tirelessly copied and collated them by myself.
When the screen is finished, people must compete to copy it.
From now on the price of paper in Nanzhong will rise.

十二年冬，微之猶滯通州，予亦未離湓上。相去萬里，不見三年，鬱鬱相念，多以吟詠自解。前後辱微之寄示之什，殆數百篇，雖藏於篋中，永以為好，不若置之座右，如見所思。由是掇律句中短小麗絕者，凡一百首，手自題錄，合為一屏。舉目會心，參若其人在於前矣。前輩作事，多出偶然。則安知此屏不為好事者所傳，異日作九江一故事爾？因題絕句，聊以獎之。

相憶采君詩作障
自書自勘不辭勞
障成定被人爭寫
從此南中紙價高[58]

As is stated in the preface, Bai Juyi copied Yuan Zhen's poems on the screen, not only to seek comfort for his own longing for Yuan, but also to create opportunities for other people to copy and spread Yuan's poem. This anticipation is full of uncertainty, as Bai considers the dissemination of literature to hinge on "contingency." Yet from the rhetorical question starting with "who knows" and from the second couplet of the poem,

"When the screen is finished, people must compete to copy it. / From now on the price of paper in Nanzhong will rise," we sense Bai's full confidence that his screen will become a new center of another round of dissemination of Yuan's work. In particular, in the poem Bai names Nanzhong as a possible destination. Nanzhong refers to a wide area in southwestern China that includes modern Yunnan, Guizhou, and southwestern Sichuan, and is traditionally associated with southern ethnic minorities that did not conform to Han Chinese culture. From Bai's perspective in Jiangzhou (Jiujiang), Tongzhou was indeed the gateway for travelers to enter Nanzhong. The evocation of this place name, therefore, has multiple implications. With this anticipation, Bai envisions that the poems will travel as far as this vast uncivilized land, and perhaps even help to civilize it by promoting poetry reading among the locals.

Read as a whole, the three poems add several new locations to the imaginary map of their literary dissemination: the Buddhist temple in Sichuan, Bai's screen in Jiangzhou, and Nanzhong, the new destination that their texts could possibly reach. The two poets were separated like floating weeds, but their painful separation ironically enabled them to schedule multiple origins and destinations for their texts across a wide range of geographical spaces. Moreover, in their work, both Yuan Zhen and Bai Juyi dramatize and mystify the uncertainties necessarily involved in the travels of their texts, which would only make their texts more charming and popular. Through copying their works, writing about copying them, and writing to anticipate prospective routes along which they would travel, the two poets actively participated in the building of an imaginary empire in which their literature dominates.

As their texts continued to travel, the empire of their poetry continued to expand. In 824, Yuan thus boasted in "Preface to Bai Juyi's Collected Works Compiled During the Changqing Reign" ("Baishi changqing ji xu" 白氏長慶集序), which he helped to edit:

> For twenty years, his [Bai Juyi's] poems have been copied on the walls of imperial palaces, temples, and post houses, and read aloud

> by aristocrats and concubines, herders and stablemen. Everywhere they are copied and duplicated, flaunted and sold, and used to barter for wine or tea.[59] In Yangzhou and Yuezhou, my and Letian's poems are often copied into books and sold in the markets. In some egregious cases, there are people who go so far as to steal our names and outrageously sell their own poems mixed together with ours. Nothing can stop it. Once I went to Pingshui Market and found that the village school children there vied with each other to learn poetry. I gathered them together to ask them what they learned. They all responded, 'The teacher taught me poems by Letian and Weizhi.' They certainly didn't know that I myself was Weizhi. There was also a merchant from Kyerim who eagerly tried to buy poems, saying that the prime minister of the kingdom often paid a gold coin per poem and could always tell the forgeries from the authentic. Since the birth of literary writings, such wide circulation has never been seen before.
>
> 二十年間，禁省、觀寺、郵堠牆壁之上無不書，王公妾婦、牛童馬走之口無不道。至於繕寫模勒，衒賣於市井，或持之以交酒茗者，處處皆是。揚越間多作書模勒樂天及予雜詩，賣於市肆之中也。其甚者，有至於盜竊名姓，苟求自售。雜亂間廁，無可奈何。予於平水市中，鏡湖傍草市名，見村校諸童競相習詩，召而問之，皆對曰："先生教我樂天、微之詩。"固亦不知予之為微之也。又雞林賈人求市頗切，自云本國宰相每以百金換一篇，其甚偽者，宰相輒能辯別之。自篇章以來，未有如是流傳之廣者。[60]

This passage celebrates the unprecedented popularity of Yuan-Bai to the point of seeming omnipresence. Their literature penetrated all the interfaces of the empire, from walls of imperial palaces to streets, villages, and markets in the peripheries, through a process of copying, exchanging, reciting, and so on. Their poetry also traveled beyond the empire to the ancient Korean kingdom, then called Kyerim. Unlike previous writings about the circulation of their work that highlight one or two individual locations, this passage provides a dazzling list of place names bracketed in the retrospective temporal frame of "for twenty years." The contour of an overarching map of literary dissemination was taking shape, and

Yuan himself was amazed by the spectacle, although he and Bai were one of the forces that made it happen.

Now when we reread the tomb inscription Bai wrote for Yuan, we realize that the place names mentioned there were not a random list but had all been mentioned in their past works: "six palaces, two capitals, and regions in eight directions" matches Yuan's passage quoted immediately before it; "the southern *man* barbarians' land" corresponds with the "Nanzhong region" that Bai envisions Yuan's poem reaching; and the eastern *yi* barbarians' kingdom refers to Kyerim. Therefore, the spatial description in Bai's tomb inscription, "From the six palaces, two capitals, and regions in eight directions to the southern *man* barbarians' land and the eastern *yi* barbarians' kingdom," not only maps out the empire of Yuan's individual and collaborative literature, but also sums up the two poets' writing about the empire.

In this way, the two poets established and delineated a verbal map that traced the dissemination of their literary works. These literary texts could "run without legs," and Yuan Zhen and Bai Juyi had a clear vision of the routes and distances they traveled and the creative mentality to engage with the journey through their various literary activities. Their demarcation of the scope of this map of dissemination corresponds with later literati's account of the popularity of their work: "From well-dressed elites to low-class commoners, they have all read and circulated [Yuan-Bai's] poems" (*Zi yiguan shizi, zhi lüyan xiali, xi chuanfeng zhi* 自衣冠士子，至閭閻下俚，悉傳諷之).[61] But what distinguished Yuan and Bai's own literary engagement with their popularity from this generic description is that they spatialized the popularity, offered it a geographic definition, and then played with this geography in interactive ways.

CONCLUSION

This chapter studies the interactive geography that Yuan Zhen and Bai Juyi established in their literary output, especially in their collaborative works.

Armed with abundant geographic knowledge, both writers demonstrated a high level of spatial consciousness in their exchange writing. In the textual space they coestablished, they enabled different places to interact with each other, which in turn embodied the interactions between two poets who had often been cut off by long distances. When Bai was in Chang'an and Yuan was stopping by Liangzhou, they kept tracing each other along their possible routes, which resulted in two poems that coincidentally synchronized the two poets in both imagination and reality. Later on, they were both demoted to the imperial southern peripheries, one in Jiangzhou and the other in Tongzhou. While the two places are separated by a daunting distance in the real world, the poets connected and integrated these locations by presenting geographical similarities in closely corresponding poetic forms. These forms included dealing with the two places in the two halves of a single couplet and writing exchange poetry in a diptych form. Finally, Yuan and Bai not only traced each other across the broad imperial space through mental mapping, but also tracked their texts as they traveled in their collaborative writing. If in the former case, the general pattern of their interactive geography was to connect the two locations that represented the two poets themselves, then in the latter case they marked out multiple locations to which their texts traveled, dramatized the traveling routes, and connected them all into a grand geographic vision of their poetic reach. Their interactive geography was an integral part of their poetic enterprise and contributed significantly to their unparalled popularity in the mid-Tang.

NOTES

1. For a recent study of the story, see Ao Wang "Poetry Matters: Interpretative Community, *Pailü* and 'Yingying Zhuan,'" in *Harvard Journal of Asiatic Studies*, vol. 71, no. 1 (June 2011): 1–34
2. As the "Biography of Yuan Zhen" 元稹傳 in *Old History of the Tang Dynasty* says: "The exceptionally intelligent and incisive Yuan Zhen became famous for his talents in his early years, and befriended Bai Juyi from Taiyuan. As a poet, Yuan was highly accomplished in describing scenery and objects. Contemporary commentators of poetry called the two 'Yuan-Bai.' Both well-dressed elites and low-class commoners read and circulated their poems written in the 'Yuanhe Style'" 稹聰警絶人，年少有才名，與太原白居易友善。工爲詩，善狀詠風態物色，當時言詩者稱元白焉。自衣冠士子，至閭閻下俚，悉傳諷之，號爲"元和體." *Jiu Tang shu*, 169.4331.
3. Due to the sheer volume of the scholarship and critical assessments, here I list only a few important general surveys and overviews in Chinese. For traditional critical opinions about Bai and Yuan, along with their collaboration, see *Gudian wenxue yanjiu ziliao huibian: Bai Juyi juan* 古典文學研究資料彙編: 白居易卷, Chen Youqin 陳友琴 ed. (Beijing: Zhonghua shuju, 1962), and *Yuan Zhen ziliao huibian* 元稹資料彙編, Yang Jun 楊軍 ed. (Beijing: Gaodeng jiaoyu chubanshe, 2014). For overviews of modern scholarship on Yuan and Bai, see Du Xiaoqin 杜曉勤, "Yuan Zhen Bai Juyi yanjiu" 元稹白居易研究, in *20 shiji Zhongguo wenxue yanjiu: Sui Tang wudai wenxue yanjiu* 二十史記中國文學研究：隋唐五代文學研究 (Beijing: Beijing chubanshe, 2001), 966–1050. For an overview of Chinese, Japanese, Korean, and Western scholarship on Yuan and Bai as well as their collaboration and joined influences in history, see Chen Caizhi 陳才智, *Yuan Bai shipai yanjiu* 元白詩派研究 (Beijing: shehui kexue wenxian chubanshe, 2007), 9–44.
4. Du Xiaoqin, "Yuan Zhen Bai Juyi yanjiu," 994.
5. For a recent discussion of the history and cultural meanings of *zhiyin*, see Anna Shields, *One Who Knows Me: Friendship and Literary Culture in Mid-Tang China* (Cambridge: Harvard University Asia Center, 2015), 47–48.

6. I use Anna Shields' translation of the long title of "Song of the Southeast." See Shields, *One Who Knows Me: Friendship and Literary Culture in Mid-Tang China*, 180.
7. Shields, *One Who Knows Me: Friendship and Literary Culture in Mid-Tang China*, 155.
8. Ibid., 142.
9. Yuan, *Yuan Zhen ji biannian jianzhu: shige juan*, 817.
10. Ibid., 818.
11. Bai, *Bai Juyi shiji jiaozhu*, 10.779.
12. Ibid., 10.777–778.
13. The Hanlin Academy, the South Palace (the Department of State Affairs), the Zhilun Gallery (the office of Secretariat) were all in the central government in Chang'an.
14. *Bai Juyi shiji jiaozhu*, 8.706.
15. Chen Yunfeng 陳允鋒, *Zhongtang wenlun yanjiu* 中唐文論研究 (Beijing: Zhongguo shehui kexue chubanshe, 2006), 248.
16. *Tang Song chuanqi zongji* 唐宋傳奇總集, Xue Hongji 薛洪勣, and Yuan Lükun 袁閭琨, eds. (Zhengzhou: Henan renmin chubanshe, 2001), 198–199.
17. Translation by Madeline Spring with modifications. See *Classical Chinese Tales of the Supernatural and the Fantastic: Selections from the Third to the Tenth Century*, ed. Karl S. Y. Kao (Bloomington: Indiana University Press, 1985), 194–195.
18. *Tang Song chuanqi zongji*, 199.
19. The first story tells "How one person dreams about going someplace and another person encounters him in reality" 彼夢有所往而此遇之也. During the reign of Empress Wu (684–740), a husband returned home and saw over the wall that there was a banquet held in a Buddhist temple. He was surprised that his wife was sitting and laughing at the banquet. The man got angry and threw a piece of tile into the banquet and broke a wine jar. All of the people inside the wall disappeared. When he returned home, his wife told him that she had a dream of exactly what he witnessed. The third story is a tale of how "two people dream the same dream" 兩相通夢. During the Zhenyuan Reign (785–805), two travelers stayed in an inn at the Tongguan Pass. One of them dreamed of himself entering the shrine of Mount Hua and having a witch pray for him. The next day he arrived at the shrine and met the witch, who had dreamed about the same thing the previous night. See *Tang Song chuanqi zongji*, 188–189.

20. Farrell, *Collaborative Circles: Friendship Dynamics and Creative Work* (Chicago: University of Chicago Press, 2001), 81, 206, 225.
21. See Xu Song 徐松, *Tang liangjing chengfang kao* 唐兩京城坊考 (Beijing: Zhonghua shuju, 1985), 4.91.
22. Many scholars have discussed how the civil service examination helped shape the literati's social networks and cultural values during the Tang. As Anthony DeBlasi says, "Although the examinations provided only a small percentage of officials serving the Tang government, other evidence makes it clear that they served an important cultural function." See DeBlasi, *Reform in the Balance: The Defense of Literary Culture in Mid-Tang China* (Albany: State University of New York Press, 2002), 47. Indeed, the civil service examination was a case of quality over quantity in terms of cultural prominence. Almost all of the important poets and writers discussed thus far in this book experienced some trajectory in the examination system, and many of them, including Han Yu and Liu Zongyuan, took the most prestigious and competitive Presented Scholar exam multiple times in order to achieve a degree.
23. For a study of the relationship between Qujiang and the literary culture of the Tang, see Li Lingfu 李令福, *Qujiang shihua* 曲江史話 (Xi'an: Sanqin chubanshe, 2005), 87–133.
24. Here is the poem: "On February Two of the Second Year of the Yongzhen Reign, the Emperor Went to the Red Phoenix Tower to Announce the Amnesty to the World. I Went on a Leisurely Walk by the Qujiang Pond with Li Gongchui [Li Shen] and Yu Shunzi [Yu Jingxiu] and Thus Missed the Grand Event.

    The spring came and I dreamt a lot, reluctant to get up in the morning, I didn't go to see the thousands of officials swarming the imperial building. A leisurely walk kept me busy. With a few friends, I stood at the headstream of the Qujiang."

    永貞二年正月二日上御丹鳳樓赦天下予與李公垂庾順之閑行曲江不及盛觀 春來饒夢慵朝起 不看千官擁御樓 却著閑行是忙事數人同傍曲江頭

    See Yuan, *Yuan Zhen ji biannian jianzhu shige juan*, 88. As the poem shows, Qujiang as a magnet place could both attract and resist. It helped Yuan and his friends rally against the power of the court when imperial politics were highly fraught. The previous year of 805 saw the transition of power from Emperor Shunzong to Emperor Xianzong, a transition shrouded with rumors about plots and conspiracies in the inner court. That year also saw the defeat of the political faction led by Wang Shuwen. Yuan and his friends in Chang'an were deeply concerned with the change of political climate and felt uneasy about what would result. Yuan's poem shows their strong reaction when politics became even more unpredictable. In 806, on the second day of the second year of the Yongzhen reign, Emperor Xianzong suddenly changed the title of his reign to Yuanhe. As a result of this change, the second year of Yongzhen Reign only had one day, and the whole country would have to change the calendar accordingly. This suspicious change signaled a dangerous instability in the imperial power. Yuan and his friends gathered at the headstream of the Qujiang Pond to distance themselves from the "thousands of officials" swarming the imperial tower to join the charade of the ceremony. This symbolic gesture against the vertical, opaque imposition of the imperial order also consolidated the integrity of Yuan's circle, which valued the active sharing of information through a fluid, person-to-person network.

25. For an overview of these poems, see Zhao Yongjian 趙永健, "Bai Juyi 'Qujiang' xilie zhong de 'dianfeng tiyan' he rensheng yiyun" 白居易"曲江"系列中的"巔峰體驗"和人生意蘊, *Zhengzhou daxue xuebao (zhexue shehui kexueban)* 鄭州大學學報 (哲學社會科學版), vol. 34, no. 5 (2001): 22–24.
26. For a comprehensive discussion of Yuan and Bai's nostalgic exchanges about their shared experiences in Chang'an and in the Qujiang area in particular, see Shields, *One Who Knows Me: Friendship and Literary Culture in Mid-Tang China*, 173–199.
27. During the Tang, the posthouse constituted a major part of the infrastructure for governmental transportation and communication, and their management required close coordination between the central government and local governments. They offered official travelers room and board and services such as exchanging horses and replenishing supplies and equipment. For an overview of the functions and cultural significance of posthouses, see Li Dehui 李德輝, *Tang Song shiqi guanyi zhidu jiqi yu wenxue zhi guanxi yanjiu* 唐宋時期館驿制度及其與文學之關系研究 (Beijing: Renmin wenxue chubanshe, 2008), 13–121.
28. See Wu Shuling, "The Development of Poetry Helped by Ancient Postal Service in the Tang Dynasty," *Frontiers of Literary Studies in China*, no. 4 (2010), 554.
29. Li, *Guben Dunhuang xiangtuzhi bazhong jianzheng*, 49–51.

30. The standard travel speed of Tang officials was seventy *li* by horse and fifty *li* by foot in one day. Ibid., 559.
31. Bai, *Waiting for the Moon: Poems of Bai Juyi*, trans. Arthur Waley (Mount Jackson, VA: Axios Press, 2012), 99–100.
32. Bai, *Bai Juyi wenji jiaozhu*, 8.360.
33. See Wu Weibin 吳偉斌, "Yuan Zhen Bai Juyi Tong Jiang change zhenxiang zongshu" 元稹白居易通江唱和真相縱述, *Nanchang daxue xuebao* 南昌大學學報, vol. 33.2 (2002.4), 100.
34. Bai, *Bai Juyi shiji jiaozhu*, 16.1245.
35. Yuan, *Yuan Zhen ji biannian jianjiao: shige juan*, 768. "Singing the ballad of Yue" as an expression means being sick and feeling homesickness. See Lu Zunwu 陸尊梧, *Gudai shici diangu cidian* 古代詩詞典故辞典 (Tianjin: Tianjin renmin chubanshe, 1992), 286.
36. *Yuan Zhen ji biannian jianjiao: shige juan*, 768.
37. Bai, *Bai Juyi shiji jiaozhu*, 16.1245.
38. *Yuan Zhen ji biannian jianjiao: shige juan*, 768.
39. Bai, *Bai Juyi shiji jiaozhu*, 16.1245.
40. Yuan, *Yuan Zhen ji biannian jianjiao: shige juan*, 768.
41. Bai, *Bai Juyi shiji jiaozhu* , 16.1245.
42. Yuan, *Yuan Zhen ji biannian jianjiao: shige juan*, 768.
43. The *you* time is the tenth double-hour of the day, from 5pm to 7pm.
44. Yuan, *Yuan Zhen ji biannian jianjiao: shige juan*, 768.
45. Liu Yuxi's "Seventy Rhymes to Express My Thoughts in Liyang," which is discussed in Chapter Three, is written in this form too.
46. See Chen Caizhi, *Yuan Bai shipai yanjiu*, 245.
47. Bai, *Bai Juyi shiji jiaozhu*, 10.818.
48. "In Response to Letian's Poems Sent to Me on His Way to Jiangzhou" 酬樂天赴江州路上見寄, *Yuan Zhen ji biannian jianzhu: shige juan*, 653.
49. Yuan, *Yuan Zhen ji biannian jianjiao: shige juan*, 768.
50. Li, *Guben Dunhuang xiangtuzhi bazhong jianzheng*, 5.
51. *Bai Juyi wenji jiaozhu*, 33.1929.
52. Yuan, *Yuan Zhen ji biannian jianzhu: shige juan*, 639.
53. Shields, *One Who Knows Me: Friendship and Literary Culture in Mid-Tang China*, 189–190.
54. Bai, *Bai Juyi shiji jiaozhu*, 15.1203–1204.
55. Shields, *One Who Knows Me: Friendship and Literary Culture in Mid-Tang China*, 190–191.
56. Yuan, *Yuan Zhen ji biannian jianzhu: shige juan*, 744.
57. Bai, *Bai Juyi shiji jiaozhu*, 17.1375.

58. *Bai Juyi shiji jiaozhu*, 17.1373–1374.
59. That Bai's poems had been used as currency was doubted by the Ming Dynasty scholar Hu Zhenheng 胡震亨 (1569–1645), but Hu later found evidence that confirmed the authenticity of Yuan's claim. As Hu says, "I had initially doubted that what Chief Minister Yuan said in his preface to Bai Juyi's *Collected Works* might not be all based on facts. Later I read in the *Fengnian lu* that 'During the Kaicheng Reign (834–848) prices were very low. To barter for some fish and meat on the village roads, ordinary men could use half a foot of rough raw silk, and the literati could use Letian's poems.' Thus what Yuan said about Bai's poems being used to barter for wine or tea had indeed happened" 初疑元相《白集序》所載未盡實，復閱《豐年錄》："開成中物價至賤，村路賣魚肉者，俗人買以胡绡半尺，士大夫買以樂天詩。"則所云交酒茗信有之. Hu, *Tangyin guiqian*, 唐音癸籤 (Shanghai: Gudian wenxue chubanshe, 1957), 25.224.
60. See *Yuan Zhen ji bianjian jianjiao: sanwen juan*, 923.
61. *Jiu Tang shu*, 169.4331.

Conclusion

Geography and Literature of the Mid-Tang

When the renowned Qing critic Yao Nai 姚鼐 (1731–1815) commented on Liu Zongyuan's landscape essay "Record of an Excursion to Huang Creek," which we discussed at length in previous chapters, he picked up a sentence that had often passed unnoticed by other critics. The sentence depicts a peculiar bird sitting on a rock by the creek: "There was a bird with a red head and black wings, as big as a swan, standing facing the east" (*Youniao chishou wuyi, daru hu, fang dongxiang li* 有鳥赤首烏翼, 大如鵠, 方東向立). Next to the expression "facing the east," Yao jotted down the following commentary:

> Master Zhu [Zhu Xi 朱熹 (1130–1200)] once pointed out that in the records of unusual things in *Classics of Mountains and Seas*, when the directions of east and west are evoked, it is because maps and pictures had appeared in the text before them. This comment is absolutely right. Liu Zongyuan did not understand this and when he wrote landscape records to imitate it, they made little sense. Later, some thought Zihou's [Liu Zongyuan] writing was well-crafted and in turn imitated him, which further missed the point."

> 朱子謂《山海經》所紀異物，有云東西嚮者，蓋以其有圖畫在前故也。此言最當。子厚不悟，作山水記效之，蓋無謂也。後人又有以子厚此等為工而效法者，益失之矣。[1]

This commentary raises two interesting points. First, Yao Nai was keenly aware that in Liu's landscape essay, Liu described the landscape as though the essay were accompanied by a map or a picture with indications of directions. In other words, what Yao discerns here is a subtle cartographical perspective embedded in Liu's landscape essay. Second, Yao's commentary shows that Liu's geography-informed writing style triggered different responses. Some admired his work and imitated his style. Others, like Yao himself, simply found it pointless. Of course, Yao's depreciation of Liu's landscape writing is in accordance with the general opinion of Liu's literature among the Tongcheng School (*Tongcheng pai* 桐城派), to which Yao belonged. Tongcheng School scholars held that Liu often imitated ancient texts blindly, leaving their traces on the surface of his own writing.[2]

Yao is not the only scholar to notice this cartographical dimension in Liu's landscape essays. Stephen Owen is also curious about the fact that Liu frequently "gives directions" in his "Eight Records of Yongzhou."[3] Unlike Yao, who quickly dismissed the feature as pointless, Owen carefully ponders the reason Liu would want to do so. As he rightly points out, writing in the southern wilderness, Liu must have presumed an audience in the northern heartland who would have had no sense of the directions in Yongzhou. In other words, "Liu Zongyuan is giving precise directions without the expectation that they will ever be used as directions." Owen then argues that this is Liu's way of orienting places in relation to other sites in Yongzhou. This, together with Liu's persistent practice of giving a name to nearly every place he visited, helped him to create order in the wilderness and make the landscape legible.[4]

Although different in approach and conclusion, both Yao's and Owen's comments point to the unusual traces of geography in Liu's literary work. But how does geography make its way into Liu's landscape essays, and perhaps into other literary representations from the same period? And what does this cross-field entanglement mean to literature, both to specific literary texts and to the literary writing of a generation of intellectuals in general? These questions constitute the major concern of my book.

My analysis has shown that Liu Zongyuan and some other great mid-Tang literary masters were far from being merely imitators of ancient geographers. Instead, they brought a broadened geographic understanding to literary creativity. This understanding was born out of their contemporary social and historical circumstances, and was part of the literati's response to the challenges they confronted in the campaign to revitalize the Tang following the destructive An Lushan Rebellion. This geographic consciousness was embedded in contemporary geographic works, and was heightened as writers engaged directly in geographic exploration during their migration and career relocation. At a time when the boundaries between fields of knowledge were porous and fluid, geography commanded wide literati interest, at times conditioning the ways the literati perceived and represented their world. An overview of the intimate historical connection between geography and literature before the mid-Tang, as well as major mid-Tang geographical achievements, can be found in chapter 1.

The majority of the book is then devoted to the question of how mid-Tang geographical advancements found entrance into the spatial imaginaries in contemporary literature, and what this cross-field entanglement means for our understanding of both fields, especially literature. My discussion of literary texts of different genres uncovers a diverse and fluid relationship between geography and literature, despite the limitations of the surviving geographical materials from the period. This cross-field affinity constituted a vital part of the epistemological and aesthetic context of the invention of spatial imaginaries in contemporaneous literature. Accordingly, a thorough exploration of this relationship gives us new clues to interpret and appreciate the literary masters' engagement with space, and to understand the mechanism of the literary writing of the time in general.

My exploration of this affinity starts with a group of map readers. In the mid-Tang, at a critical moment of imperial restoration, maps, and grand maps in particular, attained significant practical and symbolic value. No Tang maps survive today, but it is commonly known that Jia Dan's "Map of Chinese and Foreign Lands" is a monumental achievement in Chinese cartography. The map represented both the geographical objects of the Chinese empire and their corresponding celestial bodies in the heavens on a single map. Many men of letters saw the map and were amazed by its sheer scale and sophisticated representation of space. Some of these writers would internalize the worldview embodied in the map into a cartographical eye and utilize it in their literary writings about the empire, the world, and the cosmos. From Li He's "Dreaming Heaven" to Liu Zongyuan's landscape poetry and essays to Zhang Hu's poems that overlaid a grand map view onto an ancient painting, in each case, these works are imprinted with an author's eye that resembles that of a map reader.

Yet it is too much to say that the spatial imaginaries in their literature came purely from the influence of Jia's grand map. A more accurate way to put it is that the overall heightened geographical understanding and awareness at that time brought the literati a new way of looking at the world and making sense of the spaces around them. Maps, especially grand maps, gave this new spatial consciousness one form of expression; literature gave it another. These two modes of expression joined forces and afforded each other new significance. Moreover, because traditional Chinese maps are transmedia objects that combine mensural cartography with techniques of painting, religious outlooks, and textual explanations often of considerable literary value, any expression these maps offer would be inherently transmedia as well. Accordingly, the interplay between cartography and literature that we observe in the literary texts under discussion here in effect also addresses a new worldview that emerged across multiple representational modes beyond cartography and literature in the strict sense.

While grand maps conversed with literary representations of the broad space of the world or the empire, the map-guides inspired mid-Tang literary writers to probe deep into the empire's intricate local spheres. The mid-Tang literati were enthusiastic map readers, but were also diligent map-guide users, primarily out of official duty. The map-guides were the source of the authoritative regional geography in Sui and Tang. During the mid-Tang period, they were compulsory reading for newly appointed local officials as a quick guide to local affairs. With the help of the map-guides, writer-officials were able to provide a new line of spatial imaginaries that were seemingly coherent but were in fact used to offset or allude to political precariousness in the local sphere. They were coherent thanks to their large and careful selection and creative reorganization of knowledge and information from geographical references; precarious because the historical continuity found in this coherent narrative was in reality endangered for various reasons, and this is precisely the reason these writer-officials needed to write to address the issue in the first place. In this sense, the map-guides and the accumulated local knowledge they provided would become an anchor of historical continuity amidst all the ruptures and uncertainties in the post-Rebellion society. Consequently, spatial imaginaries in local writing of this kind were, in most cases, political in nature. They were composed by local officials, for purposes of local governance or political appeal to the court, and, in a broader sense, were responsive and conducive to the grand imperial restoration.

Like maps, map-guides were containers of multimedia representations of spaces, given that map-guides were composed of maps and texts. While all of these maps have been lost, the texts are still available in several hundred fragments and allusions. Our discussion focused on the map-guide is therefore primarily dedicated to the textual knowledge map-guides provided and the intertextual relationship they bore to the writer-officials' literary engagement with the local sphere. More specifically, my argument is that there was a process of what I call a cross-field multilayering of textual knowledge between geography and literature that underpinned much of the writing about localities during the mid-Tang period. To uncover this process, I worked with several literary texts that made their reference to the map-guides explicit, and when possible, I compared them to fragments from the map-guides themselves. Because citing references is largely optional in literary writing, there must have been many more works that incorporated geographical text from map-guides without explicit mention. In this sense, the works under discussion here should be understood as the tip of an iceberg, one that should indicate the frequent and fruitful encounters between geography and literature that took place during the mid-Tang period.

The mid-Tang literati's geographical understanding was acquired not only through secondary reading and learning, but also through firsthand exploration. It came not only in the form of knowledge, but also in subjective spatial experience. What best manifested the mid-Tang geographical advancements as such was the work of a group of prominent literati as they discovered the deep south amidst the mass southward immigration and development across the empire. The literary genre that most intimately expressed their southern exploration was the landscape essay, a genre representative of the time. The authors' desire to inhabit the land, a desire that resonated with waves of new immigrants in the south, made home building a central spatial imaginary in many landscape essays. Landscape essayists depicted the southern landscape from the perspective of its inhabitability, recorded their successful or failed attempts to find an ideal site for home building, and created new landmarks through building, writing, and inscribing both individually and collaboratively. The Introduction mentions the humanistic geographer Edmunds Bunkse's literary autobiography, *Geography and the Art of Life*, in which he recalls his experiences of a wide range of places, landscapes and sites to register his World War II childhood experience. In a sense we may apply Bunkse's book title to the mid-Tang landscape essayists, as their work as a whole is also marked by a combination of geographical exploration and autobiographical narrative on both the individual and collective levels.

My geographical approach contributes to existing scholarship on the landscape essay in several ways. The mid-Tang landscape essay is conventionally associated with the authors' banishment to the south and has largely been studied in the framework of banishment literature. While this approach has its merits, it has neglected the larger geographical change within which these authors were writing. The deep south was not an unchanging wilderness waiting to be discovered and poeticized, but rather was a shifting ground undergoing rapid transformation. In this context, the landscape essayists were frustrated politicians, but they were also local governors of the grand project of southern domestication. They were landscape designers, home buyers, and builders. In addition, scholars have long noticed that Liu Zongyuan's scenic depictions incorporate perspectives and expressions seen in geographical texts from earlier periods. I therefore argue that the one thing that truly sets him apart from earlier geographical writers and distinguishes his literature from geography is the detailed and delicate records of his determined efforts to inhabit and domesticate the landscape, and to make a home out of it. My approach generates a line of refreshing interpretations of some of the most celebrated landscape writing of the mid-Tang, and of Chinese literary history as a whole.

Through secondary learning and in-field explorations, the mid-Tang literati were equipped with rich geographical knowledge and diverse spatial perspectives. Yet what made them literary masters was their creative and skillful play with this geo-spatial understanding in their literary work. This can be seen in all of the writers discussed here, but perhaps most extraordinarily in the collaborative works by Yuan Zhen and Bai Juyi. An expert in cartography and an experienced traveler, respectively, Yuan and Bai brought geography into full play, turning it into articulations of personal relationships, community solidarity, and literary fame in their exchange poetry. The spatial imaginaries in their work functioned to synchronize, associate, and connect otherwise unrelated places, which reconfigured the imperial space and beyond with a literary mind's eye and consequently altered the spatial perception of their readers. I therefore call their spatial imaginaries an "interactive geography," which afforded interactions between places, separated people, and traveling texts.

Yuan and Bai's geographically informed writing unleashed the agency of literary men to the utmost. The extent of their ability to perceive geography, or the patterns of the earth, was mobilized for the purpose of literary collaboration and dissemination. If geography is essentially about humans in space, then literary geography is about creative literary minds in space. The spatial imageries in their work are never merely representations of an abstract or purely objective world, but always reflect a literary mediation of real or imaginary spaces. This mediation takes different forms in different writers. For Yuan and Bai, it is loaded with multiple layers, from complicated stylistic needs such as anadiplosis and the diptych form, to intense commitments to friendship. If we go back to the metaphor of image formation in physics that I evoked in the Introduction, the interface or the filter between the object and the image exhibits its counterpointed, kaleidoscopic characteristics in the case of Yuan and Bai, making them a revealing subject of scrutiny.

From a geographical perspective, my study brings important new insights to the already substantial scholarship on mid-Tang literature. Mid-Tang literature is an important subfield in studies of Tang literature and is known for its diversity in literary genres and themes as well as its complexity when compared to the literature of other periods in the Tang dynasty. It witnessed the diversification of poetic styles, the emergence of landscape essay, the flourishing of fantastic stories, and so on. Within each genre, different schools, trends, and personal styles competed and converged with each other at the same time.[5] For example, Stephen Owen proclaims that "in style, in topics, and in norms of treatment, the poetry of the mid-Tang is much more varied than that of the High Tang."[6] The unprecedented richness of mid-Tang literature makes it a fascinating but daunting topic for literary scholars.

My work contributes to the contextualization of this unique diversity of mid-Tang literature. First, the book highlights a close relationship between the advancement of spatial understanding and the development of literary genres, which is itself an understudied topic in Chinese literary history. The Introduction mentions that the rapid accumulation of geographical knowledge in the Han Dynasty triggered the maturation of rhapsody, the representative literary genre of the Han and the Six Dynasties. The main body of this book discusses in detail how the emergence of the landscape essay and the flourishing of office inscriptions and orisons of appeal can be accounted for by the literati's heightened geographical awareness and deepened understanding of the empire's local and frontier spaces, thanks to advancements in geography. Moreover, my book also uncovers new characteristics of specific genres. In particular, my comparative reading of literary and geographical texts highlights the literary quality of genres such as the office inscription and the orison for appeal, genres that have been commonly considered pragmatic and formulaic by literary scholars. I argue that the two genres involved careful and selective reorganization of fragmented geographical information into coherent and suggestive narratives, which often in turn empowered the author's political criticism of the central court from a deeply rooted local perspective.

Second, a geographical perspective also provides clues to our understanding of the variations within each genre and among individual writers in the mid-Tang period. This variation has been addressed throughout this book. For example, Li He has impressed both traditional and modern readers with the wild imagination and the unconventional figures and metaphors on display in his poetry; Liu Yuxi is known for incorporating local details in his poems about the south; Yuan Zhen and Bai Juyi established their collaborative literary fame through their extremely popular exchange poetry. It is as if each of these mid-Tang writers developed distinct trademarks that would distinguish them from their peers. Many historical and cultural factors contributed to this diversity, and my book shows that the advanced geographical understanding shared by the mid-Tang literati is one of them. In a time when intellectuals were also multitalented writers, the development of one field can be promptly and integrally transformed into nutrients available to other fields. In the mid-Tang, geographical advancements found entrance into contemporary individual literary realms, providing writers with fresh possibilities for writing about the world and indeed for writing the world, enabling them to claim new thematic and stylistic territories.

Finally, the first three chapters of the book start with a few literary authors who were active after the Rebellion but slightly before the mid-Tang period, namely Du Fu, Yan Zhenqing, and Yuan Jie. I incorporate them because I have identified in their writing certain patterns that anticipate the ways geography was integrated into mid-Tang literature. In literary history, the period after the Rebellion and before the mid-Tang is conventionally considered a transitional period. In terms of poetry, the Dali Reign (大曆, 766–779) of this period has long been treated as a less important but interesting bridge between the two heights of the high-Tang and the mid-Tang poetry.[7] In terms of prose, writers such as Yuan Jie from this period have been seen as forerunners of Han Yu and Liu Zongyuan's development of ancient style prose (*guwen* 古文).[8] In other words, scholars in general have treated literature of this transitional period genre by genre. My perspective on writers from this period, however, suggests that we can treat its literature as embodying a preliminary stage of the post-Rebellion epistemological development, which eventually helped allow for the blossoming of mid-Tang literature across genres.

The confluence of geography and literature is not a new discovery. On the contrary, it has attracted so much critical attention in recent decades that there is now a trend to identify components involving geography, or aspects that can be subjected to geographical analysis, in almost any literature involving space. One extreme comes from Robert T. Tally Jr., a scholar of literary cartography, who claims that "all writing partakes in a form of cartography, since even the most realistic map does not truly depict the space, but, like literature, figures it forth in a complex skein of imaginary relations."[9] Or, as Tom Conley posits, "The poet is like a cartographer insofar as it is his or her task to describe the world by mixing images, visual designs, and both aural and optical traits of language."[10] For such scholars, any literary author using poetic language to depict the world is simultaneously a geographer, and cartography has become a metaphor for a general textual-spatial exploration. In particular, for critics in a time of modernity, postmodernity, and globalization, literary works are akin to a geography in which people orient themselves in cosmopolitan cities, postindustrial landscapes, global seascapes, and various virtual realities that are endlessly labyrinthine, disorienting, and overwhelming, thus demanding a renewal of order and redemption. As James Joyce famously claimed, were Dublin to disappear, one could recreate it from his *Ulysses*.[11] In our time, the novel, as the dominant literary genre of the modern experience, has become the focal point of "literary geography," in which Charles Dickens, Herman Melville, James Joyce, Italo Calvino, and a long list of important Western writers become verbal mapmakers, traversing and reshaping geographic, social, and imaginary spaces.

Some contemporary scholars, however, have come to see this as an overemphasis on the likeness between literature and geography. As Andrew Gordon and Bernard Klein said more than a decade ago: "'Mapping,' the inscriptive practice of the cartographer, has become a key theoretical term in current critical discourse, describing a particular cognitive mode of gaining control over the world, of synthesizing cultural and geographical information, and of successfully navigating both physical and mental space. The often-inflationary usage of this suggestive semiotic concept in recent critical work tends to ignore perhaps too easily the boundaries between the metaphorical and the material."[12] Today, when it has become possible to analyze almost all elements in literature from a geographical or spatial perspective, ranging from structure and imagery in poetry to setting, character and plot in fiction, a geographical approach to literature may seem like an overused metaphor in the critical language. The poet is *like* a map-maker, which conversely makes us think that the poet is not *in fact* a map-maker. When we generally discuss a verbal projection of a landscape, a city, or any lived or imagined space, a ready-made metaphorical use of "geography" can easily slip in as a circular overgeneralization of a "likeness."

This book, in contrast, deals with the unique relationship between literature and geography in a time of radical geopolitical changes on a more substantial level than that of mere "likeness." Without denying that such "likeness" exists, I argue that it is one thing to say that literature involves spatial elements that can be approached using geographical methods, and another to say that certain written works exert their imaginative and communicative power by actively incorporating and internalizing specifically geographic knowledge, and further go on to develop and transform them into renewed figurative forms. The unique case in literary history that I present in this book is that of multitalented Tang literary masters whose geographic proficiency and interest interacted with their invention of protean forms of spatial practices and spatial imaginaries. The literary geography conducted by these poets on the page, then, does not merely constitute a metaphor, but was organically combined with actual words and activities. From their ink and their actions, new worlds flowed.

Notes

1. Yao Nai comp. Wang Wenru 王文濡 ed. *Guwen cilei zuan pingzhu* 古文辭類纂評註 (Taipei: Taiwan zhonghua shuju, 1973), 52.1.
2. For example, the representative Tongcheng School scholar Fang Bao 方苞 (1668–1749) holds a similar view on Liu Zongyuan. See Wu Wenzhi 吳文治, ed., *Liu Zongyuan ziliao huibian* 柳宗元資料彙編 (Beijing: Zhonghua shuju, 1964), vol. 2, 377–380.
3. Stephen Owen, *The End of the Chinese 'Middle Ages'*: Essays in Mid-Tang Literary Culture (Stanford: Stanford University Press, 1996), 49.
4. Ibid., 49–54.
5. For an overview of the diversity of mid-Tang literature, see Ge Xiaoyin 葛曉音, "Zhongtang wenxue de bianqian shang" 中唐文學的變遷上, *Gudian wenxue zhishi* 古典文學知識, 1994.4: 43–49, "Zhongtang wenxue de bianqian xia" 中唐文學的變遷下, *Gudian wenxue zhishi*, 1994.5: 34–39
6. Owen, *The End of the Chinese "Middle Ages": Essays in Mid-Tang Literary Culture* 2.
7. See Du, "Dali shige yanjiu" 大曆詩歌研究, in *20 shiji Zhongguo wenxue yanjiu: Sui Tang Wudai wenxue yanjiu*, 446–490.
8. See Peng Xiaole 彭小樂, "Yuan Jie de guwen gexin yu Tangdai guwen yundong" 元結的古文創新與唐代古文運動, *Zhonghua wenhua luntan* 中華文化論壇, 2015.9: 168–173.
9. See Tally Jr.'s review of Bertrand Westphal's *La Géocritique*, in *L'esprit Créateur: the International Quarterly of French and Francophone Studies*, 49.3 (Fall 2009), 134.
10. Tom Conley, *An Errant Eye: Poetry and Topography in Early Modern France* (Minneapolis: University of Minnesota Press 2011), 3.
11. *Frank Budgen, James Joyce and the Making of 'Ulysses", and Other Writings* (Oxford: Oxford University Press, 1972), 69.
12. Andrew Gordon and Bernard Klein, eds., *Literature, Mapping, and the Politics of Space in Early Modern Britain* (Cambridge: Cambridge University Press, 2010), 3.

Bibliography

Abramson, Mark. *Ethnic Identity in Tang China*. Philadelphia: University of Pennsylvania Press, 2008.

Arnold, Matthew. "The Study of Poetry." In *Classic Writings on Poetry*, edited by William Harmon, 463–484. New York: Columbia University Press, 2003.

Bai, Juyi 白居易. *Bai Juyi ji jianjiao* 白居易集箋校. Edited by Zhu Jincheng 朱金城. Shanghai: Guji chubanshe, 1988.

______. *Bai Juyi shiji jiaozhu* 白居易詩集校注. Edited and annotated by Xie Siwei 謝思煒. Beijing: Zhonghua shuju, 2006.

______. *Bai Juyi wenji jiaozhu* 白居易文集校注. Edited and annotated by Xie Siwei 謝思煒. Beijing: Zhonghua shuju, 2011.

______. *Waiting for the Moon: Poems of Bai Juyi*. Translated by Arthur Waley. Mount Jackson, VA: Axios Press, 2012.

Bi, Qiong 畢瓊 and Li Xiaocong 李孝聰. *Shanjing Shudao tu Yanjiu* 陝境蜀道圖研究. In *Ditu* 地圖, 2004.4, 45–50.

Bian, Xiaoxuan 卞孝萱, and Bian Min 卞敏. *Liu Yuxi pingzhuan* 劉禹錫評傳. Nanjing: Nanjing daxue chubanshe, 1996.

Boltz, Judith Magee. "Wuyue zhenxing tu." In *Encyclopedia of Taoism*, vol. 2, edited by Fabrizio Pregadio, 1075. London and New York: Routledge, 2001.

Bray, Francesca, Vera Dorofeeva-Lichtmann, and Georges Métailié, eds. *Graphics and Text in the Production of Technical Knowledge in China: The Warp and the Weft*. Leiden: Brill, 2007.

Brosseau, Marc. "Geography's Literature." *Progress in Human Geography* 18, no. 3 (1994): 333–353.

Budgen, Frank. *James Joyce and the Making of 'Ulysses', and Other Writings*. Oxford: Oxford University Press, 1972.

Bunkse, Edmunds Valdemars. *Geography and the Art of Life*. Baltimore, MD: Johns Hopkins University Press, 2004.

Buswell Jr., Robert E, and Donald S. Lopez Jr., eds. *The Princeton Dictionary of Buddhism*. Princeton, NJ: Princeton University Press, 2014.

Campany, Robert. *To Live as Long as Heaven and Earth: Ge Hong's Traditions of Divine Transcendents*. Berkeley: University of California Press, 2002.

Cao, Daoheng and David R. Knechtges. "Han *Fu* 賦 (*Fu* of the Han)." In *Ancient and Early Medieval Chinese Literature: A Reference Guide, Part One*, edited by David R. Knechtges and Chang Taiping, 317–333. Leiden: Brill, 2010.

Chen, Caizhi 陳才智. *Yuan Bai shipai yanjiu* 元白詩派研究. Beijing: Shehui kexue wenxian chubanshe, 2007.

Chen, Jo-shui. *Liu Tsung-yüan and Intellectual Change in T'ang China, 773–819*. New York: Cambridge University Press, 1992.

Chen, Ke 陳可. "Tangdai tingbi ji yanjiu" 唐代廳壁記研究. M.A. thesis, Shoudu shifan daxue, 2011.

Chen, Qiaoyi 陳橋驿. *Li Daoyuan pingzhuan* 酈道元評傳. Nanjing: Nanjing daxue chubanshe, 1994.

Chen, Songbai 陳松柏. "'Ming wei sima, shiwei qiutu' bian" "名為司馬, 實如囚徒" 辩. *Guangdong jishu shifan xueyuan xuebao (shehui kexue)* 廣東技術師範學院學報 (社會科學) (Jan. 2009): 17–20.

Chen, Yinke 陳寅恪. *Tangdai zhengzhishi shulun gao* 唐代政治史述論稿. Taibei: Taiwan shangwu yinshuguan, 2009.

Chen, Yixin 陳貽焮, ed. *Zhengding zhushi quan Tangshi* 增訂注釋全唐詩. Beijing: Wenhua yishu chubanshe, 2007.

Chen, Youqin 陳友琴, ed. *Gudian wenxue yanjiu ziliao huibian: Bai Juyi juan* 古典文學研究資料彙編: 白居易卷. Beijing: Zhonghua shuju, 1962.

Chen, Yuquan 陳昱全. "Bei Song yuzhi mizangquan banhua yanjiu" 北宋御制密藏詮版畫研究. M.A. thesis, National Taiwan Normal University, 2009.

Chen, Yuanlong 陳元龍, ed. *Lidai fuhui* 歷代賦匯. Bejing: Beijing tushuguan chubanshe, 1999.

Chen, Yunfeng 陳允鋒. "Lun Liu Yuxui de dili yishi jiqi shixue sixiang tedian" 論劉禹錫的地理意識及其詩學思想特點. *Ningxia shehui kexue* 寧夏社會科學 (May 2004): 106–110.

______. *Zhongtang wenlun yanjiu* 中唐文論研究. Beijing: Zhongguo shehui kexue chubanshe, 2006.

Chen, Yu-shih. *Images and Ideas in Chinese Classical Prose: Studies of Four Masters.* Stanford, CA: Stanford University Press, 1988.

Chen, Yuwen 陳毓文. "Zhong Tang wenxue de difanghua jincheng" 中唐文學的地方化進程. *Fujian sheke lutan sheke jiaoyu ban* 福建社科論壇 (社科教育版) (2007): 9–10.

Cheng, Qianfan 程千帆. *Tangdai jinshi xingjuan yu wenxue* 唐代進士行卷與文學. Shanghai: Shanghai guji chubanshe, 1980.

Clunas, Craig. *Pictures and Visuality in Early Modern China.* London: Reaktion Books Ltd., 1997.

Conley, Tom. *An Errant Eye: Poetry and Topography in Early Modern France.* Minneapolis: University of Minnesota Press 2011.

Cresswell, Tim. *Place: A Short Introduction.* Malden, MA: Blackwell, 2004.

De Weerdt, Hilde. "Maps and Memory: Readings of Cartography in Twelfth- and Thirteenth-Century Song China." *Imago Mundi* 61, no. 2 (2009): 145–167.

DeBlasi, Anthony. *Reform in the Balance: The Defense of Literary Culture in Mid-Tang China.* Albany: State University of New York Press, 2002.

Ding, Chao 丁超. "Tangdai Jia Dan de dili (ditu) zhushu jiqi dituxue chengji zaipingjia" 唐代賈耽的地理（地圖）著述及其地圖學成績再評價. In *Zhongguo lishi dili luncong* 中國歷史地理論叢 27, no. 3 (Jul. 2012): 146–156.

Ditter, Alexei. "Genre and the Transformation of Writing in Tang Dynasty China (618–907)." Ph.D. diss., Princeton University, 2009.

Dong, Gao 董誥 et al., eds. *Quan Tangwen* 全唐文. Beijing: Zhonghua shuju, 1983.

Du, Fu 杜甫. *Dushi xiangzhu* 杜詩詳注. Edited by Qiu Zhao'ao 仇兆鼇. Beijing: Zhonghua shuju, 1979.

Du, Xiaoqin 杜曉勤. *20 shiji Zhongguo wenxue yanjiu: Sui Tang wudai wenxue yanjiu* 二十史記中國文學研究：隋唐五代文學研究. Beijing: Beijing chubanshe, 2001.

Du, You 杜佑. *Tongdian* 通典. Taipei: Xinxing shuju, 1963.

Dugu, Ji 獨孤及. *Piling ji jiaozhu* 毘陵集校注. Edited by Liu Peng 劉鵬 and Li Tao 李桃. Anshan: Liaohai chubanshe, 2006.

Fan, Zhengsheng 范正聲. "Jin Shang bushi Shangguan dafu" 靳尚不是上官大夫. *Tai'an shizhuan xuebao* 泰安師專學報 (Apr. 1994): 407–408.

Fang, Zheng 方正. "Shudao yu Tangshi" 蜀道與唐詩. M.A. thesis, Xibei University, 2014.

Farrell, Michael. *Collaborative Circles: Friendship Dynamics and Creative Work.* Chicago: University of Chicago Press, 2001.

Felt, David Jonathan. "Patterns of the Earth: Writing Geography in Early Medieval China." Ph.D. diss., Stanford University, 2014.

Feng, Suiping 馮歲平. *Shanjing Shudao tu zaitan* 美國國會圖書館藏《陝境蜀道圖》再探. *Wenbo* 文博 (Feb. 2010): 33–42.

Fu, Xuancong 傅璇琮. *Tangdai keju yu wenxue* 唐代科舉與文學. Taibei: Wenshizhe chubanshe, 1994.

______ et al. *Xinbian Tang wudai wenxue biannian shi: Zhongtang juan* 新編唐五代文學編年史: 中唐卷. Shenyang: Liaohai chubanshe, 1998.

______, and Zhou Jianguo 周建國, eds. *Li Deyu wenji jianjiao* 李德裕文集校箋. Shijiazhuang: Hebei jiaoyu chubanshe, 2000.

Gao, Haifu 高海夫. *Liu Zongyuan sanlun* 柳宗元散論. Xi'an: Shanxi renmin chubanshe, 1985.

Ge, Hong 葛洪. *Shenxian zhuan jiaoshi* 神仙傳校釋. Edited by Hu Shouwei 胡守爲. Beijing: Zhonghua shuju, 2010.

Ge, Xiaoyin 葛曉音. "Zhongtang wenxue de bianqian shang" 中唐文學的變遷上, *Gudian wenxue zhishi* 古典文學知識, 1994.4: 43–49.

______, "Zhongtang wenxue de bianqian xia" 中唐文學的變遷下, *Gudian wenxue zhishi* 古典文學知識, 1994.5: 34–39

Gentzler, J. Mason. "A Literary Biography of Liu Tsung-yuan, 773–819." Ph.D. diss. Columbia University, 1966.

Gibson, James. *The Ecological Approach to Visual Perception.* New York and London: Taylor & Francis, 1986.

Gong, Yulan 龔玉蘭. *Bianzhe shiqi de Liu Zongyuan yanjiu* 貶谪時期的柳宗元研究. Nanjing: Fenghuang chubanshe, 2010.

Gordon, Andrew and Bernard Klein, eds. *Literature, Mapping, and the Politics of Space in Early Modern Britain.* Cambridge: Cambridge University Press, 2010.

Graff, David. *Medieval Chinese Warfare 300–900.* London and New York: Routledge, 2001.

Gwyther, Jordan Alexander. "Bai Juyi and the New Yuefu Movement." M.A. thesis, University of Oregon, 2013.

Han, Yu 韓愈. *Han Changli shiji biannian jianzhu* 韓昌黎詩集編年箋注. Edited by Fang Shiju 方世舉. Beijing: Zhonghua shuju, 2012.

______. *Han Changli wenji jianzhu* 韓昌黎文集校注. Collated and annotated by Ma Qichang 馬其昶. Taipei: Shijie shuju, 1967.

Harley, J. B. and David Woodward. *The History of Cartography*. Chicago: University of Chicago Press, 1994.

Hartman, Charles. *Han Yu and the T'ang Search for Unity.* Princeton, NJ: Princeton University Press, 1986.

He, Guoji 何國跻. "Liu Zongyuan de Yongzhou jiuji" 柳宗元的永州九記. *Yuwen xuexi* 語文學習 (Nov. 2008): 55–56.

He, Jie 何捷. "1200 nian qian guren liuzai daoxian de moya shike, shenyun zhijin meiyou diushi" 1200年前古人留在道县的摩崖石刻，神韵至今没有丢失. *Chenbao zhoukan*, 晨報周刊 352 (2015).

He, Ning, "Yangchuan" 洋川 (The Yang River). In Cao Yin, *Quan Tang shi* 全唐詩, vol. 735, 8400.

He, Shuzhi 何書置. "Yongzhou baji ying gaicheng yongzhou jiu ji" 永州八記應改稱永州九記. *Qiusuo* 求索 (1983): 83.

He, Wei 何畏. "Lun Nanzhong diqu de xingcheng" 論南中政區的形成. *Xingyi shifan xueyuan xuebao* 興義師範學院學報 2 (Apr. 2014): 35–39.

He, Zhuo 何焯. *Yimen dushu ji* 義門讀書記. Beijing: Zhonghua shuju, 1987.

Heaney, Seamus. *Electric Light*. London: Faber & Faber, 2001.

Hinton, David, trans. and ed. *Classical Chinese Poetry: An Anthology*. New York: Farrar, Straus and Giroux: 2010.

Hones, Sheila. *Literary Geographies: Narrative Space in* Let the Great World Spin. New York: Palgrave Macmillan, 2014).

______. *Literary Geography*. New York: Routledge, 2017.

Hou, Renzhi 侯仁之. *Zhongguo dili xue jianshi* 中國古代地理學簡史. Beijing: Kexue chubanshe, 1962.

Hu, Zhenheng 胡震亨. *Tangyin guiqian* 唐音癸籤. Shanghai: Gudian wenxue chubanshe, 1957.

Hua, Linfu 華林甫. "Sui-Tang tujing jikao shang" 隋唐《圖經》輯考（上）. *Guoli zhengzhi daxue lishi xuebao* 國#政治大學#史學報 27 (May 2007): 141–213.

______. "Sui-Tang tujing jikao xia" 隋唐《圖經》輯考（下）. *Guoli zhengzhi daxue lishi xuebao* 國立政治大學歷史學報 28 (Nov. 2007): 1–92.

Huang, Shih-shan Susan. *Picturing the True Form: Daoist Visual Culture in Traditional China*. Cambridge: Harvard University Asia Center, 2012.

Huang, Yangxing 黃陽興. "Tuxiang, yigui yu wenxue: luelun zhongtang mijiao yishu yu Han Yu de xianguai shifeng" 圖像、儀軌與文學——略論中唐密教藝術與韓愈的險怪詩風. *Wenxue yichan* 文學遺產 (Jan. 2012): 49–50.

Hucker, Charles O. *A Dictionary of Official Titles in Imperial China.* Stanford, CA: Stanford University Press, 1985.

Hung, William. *Tu Fu: China's Greatest Poet.* Cambridge, MA: Harvard University Press, 1952.

Jiang, Jianyuan 蔣見元 and Cheng Junying 程俊英, eds. *Shijing zhuxi* 詩經注析. Beijing: Zhonghua shuju, 1999.

Jao, Tsung-I 饒宗頤. *Chengxin luncui* 澄心論萃. Shanghai: Shanghai wenyi chubanshe, 1996.

Ji, Yougong 計有功. *Tangshi jishi* 唐詩紀事. Shanghai: Shanghai guji chubanshe, 1987.

Jiang, Jianyuan 蔣見元 and Cheng Junýing 程俊英, eds. *Shijing zhuxi* 詩經注析. Beijing: Zhonghua shuju, 1999.

Jiang, Peng 江鵬. "Cangwu rushi kao" 蒼梧入詩考. *Wuzhou xueyuan xuebao* 梧州學院學報 2, no. 24 (Feb. 2014): 48–55.

Jiang, Xinhong 蔣新紅. "Liu Zongyuan 'Yongzhou baji' meixue meili de xiezuo xue jiegou" 柳宗元《永州九記》美學魅力的寫作學解構. *Guangdong Peizheng xueyuan xuebao* 广东培正学院学报 7, no. 3 (Sept. 2007): 61–64.

Kao, Karl S. Y., ed. *Classical Chinese Tales of the Supernatural and the Fantastic: Selections from the Third to the Tenth Century.* Bloomington: Indiana University Press, 1985.

Ke, Wancheng 柯萬成. "Han Yu bianchao xingji yu sanshi jinian xinlun" 韓愈貶潮行跡與三詩繫年新論. In *Han Yu yu Lingnan wenhua* 韓愈與嶺南文化, edited by Zhang Qinghua 張清華, Hu Axiang 胡阿祥, and Liu Yingjie 劉英傑, 370–386. Beijing: Xueyuan chubanshe, 2002.

Knechtges, David R., and Taiping Chang, eds. *Ancient and Early Medieval Chinese Literature: A Reference Guide,* vol. 3 & 4. Leiden: Brill Academic Publishers, 2013.

______, trans. *Wenxuan, or Selections of Refined Literature.* Princeton, NJ: Princeton University Press, 1996.

Kohn, Livia, ed. *Daoism Handbook.* Leiden: Brill, 2000.

Kroll, Paul. "Lexical Landscapes and Textual Mountains in the High Tang." *T'oung Pao,* Second Series 84 (1998): 62–101.

______, ed. *A Student's Dictionary of Classical and Medieval Chinese.* Leiden: Brill, 2014.

Lai, Xinxia 來新夏. *Zhongguo difangzhi* 中國地方志. Taipei: Shangwu yinshuguan, 1995.

Lando, Fabio. "Fact and Fiction: Geography and Literature: A Bibliographic Survey." *GeoJournal* 38, no. 1 (January 1996): 3–18.

Laozi 老子. *Laozi Daode jing* 老子道德經. Annotated by Wang Bi 王弼. Beijing: Zhonghua shuju, 1985.

James Legge, ed. and trans. *The Chinese Classics.* London, 1865.

Leggett, B. J. "Stevens' Late Poetry." In *The Cambridge Companion to Wallace Stevens,* edited by John N. Serio. Cambridge: Cambridge University Press, 2007.

Lei, Wen 雷聞. *Jiaomiao zhiwai—Sui Tang guojia jisi yu zongjiao* 郊庙之外——隋唐国家祭祀与宗教. Beijing: Sanlian shudian, 2009.

Lewis, Mark Edward. *China's Cosmopolitan Empire: The Tang Dynasty.* Cambridge: Belknap Press of Harvard University Press, 2009.

______. *The Construction of Space in Early China.* Albany: State University of New York Press, 2006.

______. *The Flood Myths of Early China.* Albany: State University of New York Press, 2006.

Li, Dehui 李德輝. *Tang Song shiqi guanyi zhidu jiqi yu wenxue zhi guanxi yanjiu* 唐宋時期館驿制度及其與文學之關系研究. Beijing: Renmin wenxue chubanshe, 2008.

Li, Deyu. "Memorial to present the *Xi'nan beibian lu*" 進《西南備邊錄》狀. In *Li Deyu wenji jianjiao* 李德裕文集校箋, edited by Fu Xuancong 傅璇琮 and Zhou Jianguo 周建國, 351. Shijiazhuang: Hebei jiaoyu chubanshe, 2000.

Li, He李賀. *Li Changji geshi biannian jianzhu* 李長吉歌詩編年箋注. Edited and annotated by Wu Qiming 吳企明. Beijing: Zhonghua shuju, 2012.

Li, Ho. *The Poems of Li Ho.* Translated by J. D. Frodsham. Oxford: Clarendon Press, 1970.

Li, Jifu 李吉甫. *Yuanhe junxian tuzhi* 元和郡縣圖志. Beijing: Zhonghua shuju, 1983.

Li, Lingfu 李令福. *Qujiang shihua* 曲江史話. Xi'an: Sanqin chubanshe, 2005.

Li, Weishi 李偉實. "Shangguan dafu he Jin Shang shi liangge ren" 上官大夫和靳尚是兩個人. *Linglin shifan gaodeng zhuanke xuexiao xuebao* 零陵師範高等專科學校學報 (Feb. 1999): 33–34.

Li, Zhao 李肇. *Tang guoshi bu* 唐國史補. Shanghai: Shanghai guji chubanshe, 1979.

Li, Zhengyu 李正宇. *Guben Dunhuang xiangtuzhi bazhong jianzheng* 古本敦煌鄉土志八種箋證. Taibei: Xinwenfeng chubanshe, 1998.

______. *Guben Dunhuang xiangtuzhi bazhong jianzheng* 古本敦煌鄉土志八種箋證. Lanzhou: Gansu renmin chubanshe, 2008.

Liang, Qichao 梁啟超. "Zhongguo dili dashi lun." In *Yinbingshi wenji juan 14* 飲冰室文集卷十四, 19–38. Shanghai: Guangzhi shuju, 1902.

Liao, Yifang 廖宜方. *Tangdai de lishi jiyi* 唐代的歷史記憶. Taipei: Taida chuban zhongxin, 2010.

Liu, An 劉安 et al. *Huainanzi jishi* 淮南子集釋. Edited by He Ning 何寧. Beijing: Zhonghua shuju, 1998.

Liu, Guoying 劉國盈. *Han Yu pingzhuan* 韓愈評傳. Beijing: Beijing shifan xueyuan chubanshe, 1991.

Liu, Xingchao 劉興超. "Lun Tangdai tingbi ji" 論唐代廳壁記. *Sichuan daxue xuebao (zhexue shehui kexue ban)* 四川大學學報（哲學社會科學版）3 (2008): 133–137.

Liu, Xu 劉昫 et al. *Jiu Tang shu*. Beijing: Zhonghua shuju, 1975.

Liu, Yuxi 劉禹錫. *Liu Yuxi quanji biannian jiaozhu* 劉禹錫全集編年校注. Edited by Tao Min 陶敏 and Tao Hongyu 陶紅雨. Changsha: Yuelu shushe, 2003.

Liu, Zhangqing 劉長卿. *Liu Zhangqing shi biannian jianzhu* 劉長卿詩編年箋注. Edited and annotated by Chu Zhongjun 儲仲君. Beijing: Zhonghua shuju, 1995.

Liu, Zhenlun 劉真倫, and Yue Zhen 岳珍, eds. *Han Yu wenji huijiao jianzhu* 韓愈文集匯校箋注. Beijing: Zhonghua shuju, 2010.

Liu, Zongyuan 柳宗元. *Liu Zongyuan shi jianshi* 柳宗元詩箋釋. Annotated by Wang Guo'an 王國安. Shanghai: Shanghai guji chubanshe, 1998.

______. *Liu Zongyuan ji jiaozhu* 柳宗元集校注. Edited and annotated by Yin Zhanhua 尹占華 and Han Wenqi 韓文奇. Beijing: Zhonghua shuju, 2013.

Lü, Dafang 吕大防. *Han Yu nianpu* 韩愈年谱. Beijing: Zhonghua shuju, 1991.

Lu, Xinyuan 陸心源, ed. *Tang wen shiyi* 唐文拾遺. Taipei: Wenhai chubanshe, 1962.

Lu, Zunwu 陸尊梧. *Gudai shici diangu cidian* 古代詩詞典故辞典. Tianjin: Tianjin renmin chubanshe, 1992.

Luo, Zhenyu 羅振玉, ed. *Mingsha shishi yishu zhengxubian* 鳴沙石室佚書正續編. Beijing: Beijing tushu chubanshe, 2004.

Marks, Robert B. *Tigers, Rice, Silk, and Silt: Environment and Economy.* Cambridge: University of Cambridge Press, 1998.

McMullen, David. "Bureaucrats and Cosmology: The Ritual Code of T'ang China." In *Rituals of Royalty: Power and Ceremonial in Tradi-*

tional Societies, edited by David Cannadine and Simon Price, 181–236. Cambridge: Cambridge University Press, 1987.

McNair, Amy. *The Upright Brush: Yan Zhenqing's Calligraphy and Song Literati Politics.* Honolulu: University of Hawai'i Press, 1998.

Miller, Peter N., and François Louis. *Antiquarianism and Intellectual Life in Europe and China, 1500–1800.* Ann Arbor: University of Michigan Press, 2012.

Mitchell, W. J. T. "What Is an Image." In *Iconology: Images, Text, Ideology,* 7–46. Chicago: The University of Chicago Press, 1986.

Needham, Joseph. *Science and Civilization in China.* Cambridge: Cambridge University Press, 1954–2008.

Nienhauser, William, ed. *The Indiana Companion to Tradtional Chinese Literature.* Bloomington: Indiana University Press, 1986.

______, ed. *Tang Dynasty Tales: A Guided Reader.* Singapore; Hackensack, NJ: World Scientific, 2010.

______, Charles Hartman, William Bruce Crawford, Jan Walls, and Lloyd Neighors. *Liu Tsung-yüan.* New York: Twayne Publishers Inc., 1973.

Ooms, Herman. *Imperial Politics and Symbolics in Ancient Japan: The Tenmu Dynasty, 650–800.* Honolulu: University of Hawai'i Press, 2009.

Ouyang, Xiu 歐陽修. 歐陽修全集. Xianggang: Guangzhi shuju, 1970.

Owen, Stephen. "The Cultural Tang." In *The Cambridge History of Chinese Literature,* vol. 1, edited by Kang-i Sun Chang and Stephen Owen, 286–380. Cambridge: Cambridge University Press, 2013.

_______. *The Late Tang: Chinese Poetry of the Mid-Ninth Century (827–860).* Cambridge, MA: Harvard University Asia Center, 2006.

_______, trans. *The Poetry of Du Fu.* Boston: De Gruyte, 2016.

_______. *The Poetry of Meng Chiao and Han Yü.* New Haven, CT: Yale University Press, 1975.

_______. *The End of the Chinese "Middle Ages:" Essays in Mid-Tang Literary Culture.* Stanford: Stanford UP, 1996.

Pan, Sheng 潘晟. *Ditu de zuozhe jiqi yuedu: yi Song-Ming wei hexin de zhishishi kaocha* 地圖的作者及其閱讀：以宋明為核心的知識史考察. Nanjing: Jiangsu renmin chubanshe, 2013.

Pankenier, David. *Astrology and Cosmology in Early China.* Cambridge: Cambridge University Press, 2013.

Park, Hyunhee. *Mapping the Chinese and Islamic Worlds: Cross-cultural Exchange in Pre-modern Asia.* Cambridge: Cambridge University Press, 2012.

Pelliot, Paul. "Le 'Cha-tcheou-tou-fou-t'ou-king' et la colonie sogdienne de la region du Lob Nor." *Journal Asiatique* (1916): 111–123.

Peng, Dingqiu 彭定求 et al., eds. *Quan Tangshi* 全唐詩. Beijing: Zhonghua shuju, 1960.

Peng, Min 彭敏. "Yuan Jie 'Cishan mingxu' chutan" 元結《次山銘敘》初探. *Hunan keji xueyuan xuebao* 湖南科技學院學報 32, no. 5 (May 2011): 31–33.

Peng, Xiaole 彭小樂. "Yuan Jie de guwen gexin yu Tangdai guwen yundong" 元結的古文創新與唐代古文運動, *Zhonghua wenhua luntan* 中華文化論壇, 2015.9: 168–173.

Pocock, Douglas C.D., ed. *Humanistic Geography and Literature (RLE Social & Cultural Geography): Essays on the Experience of Place.* London: Croom Helm, 1981.

Qi, Hehui 祁和暉. "Dufu Qinzhou shi ji xie xiyu sichou zhilu shouduan zhancheng shanchuan renwen fengmao" 杜甫秦州詩記寫西域絲綢之路首段棧程山川人文風貌. *Du Fu yanjiu xuekan* 杜甫研究學刊 4 (2014): 16–21.

Qian, Nanxiu. "Liu Zongyuan." In *Encyclopedia of the Essay*, edited by Tracy Chevalier, 493–494. London and Chicago: Fitzroy Dearborn Publishers, 1997.

Qu, Xinfu 屈新福. "Li Jifu jiqi Yuanhe junxian tuzhi" 李吉甫及其元和郡縣志. M.A. thesis, Anhui University, 2009.

Reed, Carrie. *A Tang Miscellany: An Introduction to Youyang Zazu.* New York: Peter Lang Publishing, Inc., 2003.

Ren, Fang 任昉. *Shuyi ji* 述異記. Beijing: Zhonghua shuju, 1960.

Ruan, Weizhou 阮維周. *Zhonghua dizhi xueshi* 中國地質學史. Taipei: Shangwu yinshuguan, 1985.

van Schaik, Sam. *Tibet: A History.* New Haven: Yale University Press, 2013.

Seamon, David. "Reconciling Old and New Worlds: The Dwelling–Journey Relationship as Portrayed in Vilhelm Moberg's 'Emigrant' Novels." In *Dwelling, Place and Environment: Towards a Phenomenology of Person and World*, edited by David Seamon and Robert Mugerauer, 227–245. Boston: Martinus Nijhoff Publishers, 1985.

Seidel, Anna. *Imperial Treasures and Taoist Sacraments; Taoist Roots in the Apocrypha.* In *Tantric and Taoist Studies in Honour of R. A. Stein*, vol. 2, edited by Michel Strickmann. Bruxelles: Inst. Belge des Hautes Études Chinoises, 1983.

Shang, Yongliang 尚永亮. *Bianzhe wenhua yu bianzhe wenxue: yi Zhong Tang Yuan he wuda shiren zhibian jiqi chuangzuo wei zhongxin* 貶謫文化与貶謫文學：以中唐元和五大詩人之貶及其創作为中心. Lanzhou: Lanzhou daxue chubanshe, 2003.

Sharp, Joanne P. "Towards a Critical Analysis of Fictive Geographies." *Area* 32, no. 3 (September 2000): 327–334.

Shen, Deqian 沈德潛. *Tang Song badajia duben* 唐宋八大家讀本, vol. 9. Hefei: Anhui wenyi chubanshe, 1998.

Shen, Zengzhi 沈增植. *Hairi lou zhacong* 海日樓劄叢. Beijing: Zhonghua shuju, 1963.

Shi, Shuo 石碩. "Han Jin shiqi Nanzhong yi yue bianxi" 漢晉時期南中夷、越辨析. *Minzu yanjiu* 民族研究 (Jan. 2006): 59–66.

Shi, Ziyu 施子瑜. *Liu Zongyuan nianpu* 柳宗元年譜. Wuhan: Hubei renmin chubanshe, 1958.

Shields, Anna. *One Who Knows Me: Friendship and Literary Culture in Mid-Tang China.* Cambridge, MA: Harvard University Asia Center, 2015.

Shimizu, Shigeru 清水茂. "Yū Sōgen no seikatsu taiken to sono sansuiki" 柳宗元の生活体験とその山水記. *Chūgoku bungakuhō* 中國文學報 (1955-04): 45–74.

Shimosada, Masahiro 下定雅弘. *Ryū Sōgen: gyakkyō o ikinuita utsukushiki tamashii* 柳宗元: 逆境を生きぬいた美しき魂. Tokyo: Bensei Shuppan, 2009.

Shizunaka, Takeshi 静永健. *Kanseki denrai: Haku Rakuten no shiika to Nihon* 漢籍伝来: 白楽天の詩歌と日本. Tokyo: Bensei Shuppan, 2010.

William Shurtleff, H.T. Huang, and Akiko Aoyagi. *History of Soybeans and Soyfoods in China and Taiwan, and in Chinese Cookbooks, Restaurants, and Chinese Work with Soyfoods outside China: Extensively Annotated Bibliography and Sourcebook including Manchuria, Hong Kong and Tibet (1024 BCE to 2014)*. Lafayette, CA: Soyinfo Center, 2014.

Sivin, Nathan and Gari Ledyard. "Introduction to East Asian Cartography." Harley and Woodward, *The History of Cartography*, 23–31.

Sima, Qian 司馬遷. *Records of the Grand Historian of China: Translated from The Shih Chi (Volume II: The Age of Emperor Wu 140 to Circa 100 B.C.)*. Translated by Burton Watson. New York: Columbia University Press, 1971.

______. *Shiji* 史記. Beijing: Zhonghua shuju, 1959.

Smith, Richard. *Chinese Maps: Images of "All Under Heaven."* Oxford: Oxford University Press, 1996.

______. *Mapping China and Managing the World: Culture, Cartography and Cosmology in Late Imperial Times*. London and New York: Routledge, 2013.

Song, Xi 宋晞. *Fangzhi xue yanjiu luncong* 方志學研究論叢. Taipei: Taiwan shangwu yinshuguan, 1999.

Spring, Madeline K. "T'ang Landscapes of Exile." *Journal of the American Oriental Society* 117, no. 2 (Apr. – Jun. 1997): 312–323.

Strassberg, Richard E. *Inscribed Landscapes: Travel Writing from Imperial China*. Berkeley: University of California Press, 1996.

Tackett, Nicolas. *The Destruction of Chinese Medieval Aristocracy.* Cambridge, MA: Harvard University Asia Center, 2016.

Tally Jr., Robert. *Melville, Mapping, and Globalization: Literary Cartography in the American Baroque Writer.* London: Continuum, 2009.

______. "Review of Bertrand Westphal's *La Géocritique.*" *L'esprit Créateur: The International Quarterly of French and Francophone Studies* 49, no. 3 (Fall 2009): 134.

______. *Spatiality.* New York: Routledge, 2013.

Tian, Feng 田峰. "Du Fu cong Qinzhou dao Ba Shu Jing Xiang de dili ganzhi yu wenhua tiyan" 杜甫從秦州到巴蜀荊湘的地理感知與文化體驗. *Zhongguo yunwen xuekan* 中國韻文學刊 30, no.1 (Jan. 2016): 11–17.

Tian, Xiaofei. "Literary Learning: Encyclopedias and Epitomes." In *The Oxford Handbook of Classical Chinese Literature (1000 BCE–900 CE),* edited by Wiebke Denecke, Wai-Yee Li, and Xiaofei Tian, 132–146. New York: Oxford University Press, 2017.

______. *Visionary Journeys: Travel Writings from Early Medieval and Nineteenth-Century China.* Cambridge, MA: Harvard University Asia Center, 2012.

Tu, Kuo-Ching. *Li Ho.* Boston: Twayne Publishers, 1979.

Tuan, Yi-fu. "Space and Place: Humanistic Perspective." In *Progress in Geography,* vol. 6, edited by C. Board, R.J. Chorley, P. Haggett, and D.R. Stoddart, 211–252. London: E. Arnold, 1974.

______. *Space and Place: The Perspective of Experience.* Minneapolis: University of Minnesota Press, 2001. First published 1977.

Twitchett, Denis C., ed. *The Cambridge History of China,* vol. 3: *Sui and Tang China, 589–906 AD, Part 1.* Cambridge: Cambridge University Press, 1979.

Unwin, Tim. "A Waste of Space? Towards a Critique of the Social Production of Space." *Transactions of the Institute of British Geographers* 25, no. 1 (2000): 11–29.

Varsano, Paula. "Do You See What I See?: Visuality and the Formation of the Chinese Landscape." *Chinese Literature: Essays, Articles, Reviews* 35 (2013): 31–57.

Waley, Arthur, trans., with additional translations by Joseph R. Allen. *The Book of Songs.* New York: Grove Press, 1996.

______. *The Life and Times of Po Chü-I.* London: George Allen & Unwin, 1951.

______, trans. *Waiting for the Moon: Poems of Bai Juyi.* Edinburg, VA: Axios Press, 2012.

Wang, Ao. "Poetry Matters: Interpretative Community, *Pailü* and 'Yingying Zhuan.'" *Harvard Journal of Asiatic Studies* 71, no. 1 (June 2011): 1–34.

Wang, Chengzu 王成組. *Zhongguo dilixue shi: Xianqin zhi mingdai* 中國地理學史: 先秦至明代. Beijing: Shangwu yinshuguan, 2005.

Wang, Chongmin王重民. *Dunhuang guji xulu* 敦煌古籍敘錄. Beijing: Zhonghua shuju, 1979.

Wang, Guoying 王國瓔. *Zhongguo wenxue shi xinjiang* 中國文學史新講. Taipei: Lianjing chuban gongsi, 2014.

Wang, Jian 王建. *Wang Jian shiji jiaozhu* 王建詩集校注. Edited and annotated by Yin Zhanhua 尹占華. Chengdu: Bashu shushe, 2006.

Wang, Liqun 王立群. *Zhongguo gudai shanshui youji yanjiu* 中國古代山水遊記研究. Beijing: Zhongguo shehui kexue chubanshe, 2008.

Wang, Pu 王溥. *Tang hui yao* 唐會要. Beijing: Zhonghua shuju, 1955.

Wang, Qianjin 汪前進. "Xiancun zui wanzheng de yifen Tangdai dili quantu shuju ji" 現存最完整的一份唐代地理全圖數據集. *Ziran kexue shi yanjiu* 自然科學史研究17, no. 3 (1998): 273–288.

Wang, Xiangzhi 王象之. *Yudi jisheng* 輿地紀勝, vol. 5. Hangzhou: Zhejiang guji chubanshe, 2012.

Wei, Zheng 魏征 et al. *Sui shu* 隋書. Beijing: Zhonghua shuju, 1973.

Wu, Heqing 吳河清. "Lun Tangdai tingbi ji de wenxian jiazhi" 論唐代廳壁記的文獻價值, *Henan daxue xuebao (shehui kexue ban)* 河南大學學報（社會科學版）51, no. 3 (May 2011): 108–113.

Wu, Songdi 吳松弟. *Zhongguo yimin shi* 中國移民史vol. 3. Fuzhou: Fujian renmin chubanshe, 1997.

Wu, Shuling. "The Development of Poetry Helped by Ancient Postal Service in the Tang Dynasty." *Frontiers of Literary Studies in China* 4 (2010): 553–577.

Wu, Weibin 吳偉斌. "Yuan Zhen Bai Juyi Tong Jiang changhe zhenxiang zongshu" 元稹白居易通江唱和真相縱述. *Nanchang daxue xuebao* 南昌大學學報 33, no. 2 (Apr. 2002): 97–102.

Wu, Wenzhi 吳文治, and Xie Hanqiang 謝漢強, eds. *Liu Zongyuan da cidian* 柳宗元大辭典. Hefei: Huangshan shushe, 2004.

Wu, Xiaolin 吳小林. *Liu Zongyuan de sanwen yishu* 柳宗元的散文藝術. Taiyuan: Shanxi renmin chubanshe, 1989.

Wu, Zaiqing 吳在慶, and Yang Juanjuan 楊娟娟. "Han Yu bianzhe Yangshan yuan yin kaoxi" 韩愈贬阳山原因考析. *Zhongzhou xuekan* 中州學刊 (2006): 182–195.

Xiao, Chi 蕭馳. "Cong shanshui dao shui shi: Yuan Jie, Liu Zongyuan yu Zhongtang shanshui meigan huayu de yizhong bianhua" 從「山水」到「水石」： 元結、柳宗元與中唐山水美感話語的一種變化. *Zhongzheng hanxue yanjiu* 中正漢學研究 (December 2014): 293–332.

Xiao, Tong 蕭統, ed. *Wenxuan* 文選. Shanghai: Shanghai guji chubanshe, 1986.

Xie, Lingyun 謝靈運. "Written in a Stone Study Surrounded by the Lake" 石壁精舍還湖中作. In *Xie Kangle shizhu* 謝康樂詩注, 63. Beijing: Renmin wenxue chubanshe, 1958.

Xin, Deyong 辛德勇. "Shuo fuchang shike Yuji tu yu Huayi tu" 說阜昌石刻《禹跡圖》與《華夷圖》. *Yanjing xuebao* 燕京學報 28 (2010): 1–72.

______. "Tangdai de dilixue" 唐代的地理學. In *Tangdai Diyu jiegou yu yunzuo kongjian* 唐代地域結構與運作空間, edited by Li Xiaocong 李孝聰, 439–463. Shanghai: Shanghai cishu chubanshe, 2003.

Xu, Song 徐松. *Tang liangjing chengfang kao* 唐兩京城坊考. Beijing: Zhonghua shuju, 1985.

Xue, Hongji 薛洪勣, and Yuan Lükun 袁閭琨, eds. *Tang Song chuanqi zongji* 唐宋傳奇總集. Zhengzhou: Henan renmin chubanshe, 2001.

Yao, Nai 姚鼐 comp., Wang Wenru 王文濡 ed. *Guwen cilei zuan pingzhu* 古文辭類纂評註. Taipei: Taiwan zhonghua shuju, 1973.

Yan, Chunhua 嚴春華. "Tangdai jishen wen lunlüe" 唐代祭神文論略. *Hengyang shifan xueyuan xuebao* 衡陽師範學院學報33, no. 2 (Apr. 2012): 79–83.

Yang, Chengzu 楊承祖. *Yuan Jie yanjiu* 元結研究. Taibei: Guoli bianyiguan, 2002.

Yang, Jun 楊軍, ed. *Yuan Zhen ziliao huibian* 元稹資料彙編. Beijing: Gaodeng jiaoyu chubanshe, 2014.

Yang, Xiaoshan. "Naming and Meaning in the Landscape Essays of Yuan Jie and Liu Zongyuan." *Journal of the American Oriental Society* 120, no. 1 (Jan.–Mar. 2000): 82–96.

Yang, Yi 楊義. *Wenxue dilixue huitong* 文學地理學會通. Beijing: Zhongguo shehui kexue chubanshe, 2013.

Yang, Yuejun 楊月君. *Tangdai jingji diqu zhian guanli yanjiu* 唐代京畿地區治安管理研究. Beijing: Zhongguo shehui kexue chubanshe, 2009.

Yang, Ziyi 楊子怡. *Han Yu cichao yu Su Shi yuhui bijiao yanjiu* 韓愈刺潮與蘇軾寓惠比較研究. Chengdu: Bashu shushe, 2008.

Yee, Cordell D. K. "Cartography in China." In Harley and Woodward, *The History of Cartography*, 34–231.

Yin, Zhanhua 尹占華. Preface to *Zhang Hu shiji jiaozhu* 張祜詩集校注, Zhang Hu, 1–29. Chengdu: Bashu shushe, 2007.

Yu, Gang 俞鋼. *Tangdai wenyan xiaoshuo yu keju zhidu* 唐代文言說與科舉制度. Shanghai: Shanghai guji chubanshe, 2004.

Yu, Shinan 虞世南, comp. *Beitang shuchao* 北堂書鈔. Tianjin: Tianjin guji chubanshe, 1988.

Yuan, Jie 元結. *Yuan Cishan ji* 元次山集. Shanghai: Zhonghua shuju, 1927.

Yuan, Zhen 元稹. *Yuan Zhen ji biannian jianzhu: shige juan* 元稹集編年箋註：詩歌卷. Edited and annotated by Yang Jun 楊軍. Xi'an: Sanqin chubanshe, 2002.

______. *Yuan Zhen ji biannian jianjiao: sanwen juan* 元稹集編年箋校：散文卷. Edited and annotated by Yang Jun 楊軍. Xi'an: Sanqin chubanshe, 2008.

Zeng, Daxing 曾大興. *Wenxue dilixue yanjiu* 文學地理學研究. Beijing: Shangwu yinshuguan, 2012.

Zhang, Hu 張祜. *Zhang Hu shiji jiaozhu* 張祜詩集校注. Edited and annotated by Yin Zhanhua 尹占華. Chengdu: Bashu shushe, 2007.

Zhang, Ji 张籍. *Zhang Ji ji xinian jiaozhu* 張籍集繫年校注. Beijing: Zhonghua shuju, 2011.

Zhang, Jinguang 張晉光. *Anshi zhi luan dui Tangdai jingji fazhan yingxiang yanjiu* 安史之亂對唐代經濟發展影響研究. Beijing: Zhongguo caizheng jingji chubanshe, 2008.

Zhang, Qizhi 張豈之, Zhang Guogang 張國剛, and Yang Shusen 楊樹森, eds. *Zhongguo lishi: Sui Tang Song shi* 中國歷史—隋唐宋史. Taibei: Wunan tushu gongsi, 2002.

Zhang, Peiheng 章培恒, and Luo Yuming 駱玉明, eds. *Zhongguo wenxueshi* 中國文學史. Shanghai: Fudan daxue chubanshe, 1999.

Zhang, Shizhao 章士釗. *Liuwen zhiyao* 柳文指要. Beijing: Zhonghua shuju, 1971.

Zhang, Weiran 張偉然. *Zhonggu wenxue de dili yixiang* 中古文學的地理意象. Beijing: Zhonghua shuju, 2014.

Zhang, Zhongcai 張仲裁. *Tang Wudai wenren rushu kaolun* 唐五代文人入蜀考. Beijing: Shehui kexue chubanshe, 2013.

Zhao, Yongjian 趙永健. "Bai Juyi 'Qujiang' xilie zhong de 'dianfeng tiyan' he rensheng yiyun" 白居易"曲江"系列中的"巔峰體驗"和人生意蘊. *Zhengzhou daxue xuebao (zhexue shehui kexueban)* 鄭州大學學報 (哲學社會科學版) 34, no. 5 (2001): 22–24.

Zhao, Yuanyi 趙元一. *Fengtian lu* 奉天錄. Beijing: Zhonghua shuju, 1985.

Zhao, Zhen 趙貞. "Lun Tangdai tujing de bianxiu" 論唐代圖經的編修. *Shixuan shi yanjiu* 史學史研究 (Apr. 2013): 88–98.

Zhou, Shaoliang 周紹良."Du Shazhou tujing juan zi" 讀《沙州圖經》卷子. *Dunhuang yanjiu* 敦煌研究 2 (1987): 27–33.

Zhou, Xianglu. *Yuan Zhen nianpu xinbian* 元稹年譜新編. Shanghai: Shanghai guji chubanshe, 2004.

Zhou, Zhenhe 周振鶴. *Xuela yishijiu* 學臘一十九. Jinan: Shandong jiaoyu chubanshe, 1999.

Zhu, Guantian 朱關田. *Yan Zhenqing nianpu* 顏真卿年譜. Hangzhou: Xiling yinshe chubanshe, 2008.

Zou, Jianjun 鄒建軍. "Wenxue dilixue yanjiu de zhuyao lingyu" 文學地理學研究的主要領域. *Shijie wenxue pinglun* 世界文學評論 (Jan. 2009): 41–46.

Index

Xiang River, 203, 209–210, 219–223, 230, 232, 246

Xiao River, 209, 246

Xie Lingyun, 137, 139–140, 227, 238

About the Author

Ao Wang is an Associate Professor of Chinese Literature at Wesleyan University. Dr. Wang holds a PhD from Yale University and an MA from Washington University in St. Louis. He has published five books of his own poetry and is the recipient of the Anne Kao Poetry Prize and the New Poet Prize from China's "Shi Jianshe" Poetry Society.

Cambria Sinophone World Series

General Editor: Victor H. Mair (University of Pennsylvania)

Books in the Cambria Sinophone World Series

Spatial Imaginaries in Mid-Tang China: Geography, Cartography, and Literature by Ao Wang

Texts and Transformations: Essays in Honor of the 75th Birthday of Victor H. Mair edited by Haun Saussy

Chinese Women Writers and Modern Print Culture by Megan M. Ferry

Reading Lu Xun Through Carl Jung by Carolyn T. Brown

Gao Xingjian and Transmedia Aesthetics edited by Mabel Lee and Liu Jianmei

Imperfect Understanding: Intimate Portraits of Chinese Celebrities edited by Christopher Rea

Zhang Yimou: Globalization and the Subject of Culture by Wendy Larson

The Borderlands of Asia: Culture, Place, Poetry by Mark Bender

Buddhist Transformations and Interactions: Essays in Honor of Antonino Forte edited by Victor H. Mair

Chinese Avant-garde Fiction: Quest for Historicity and Transcendent Truth by Zhansui Yu

Eroticism and Other Literary Conventions in Chinese Literature: Intertextuality In The Story Of The Stone by I-Hsien Wu

The Sinophone Cinema of Hou Hsiao-hsien: Culture, Style, Voice and Motion by Christopher Lupke

Supernatural Sinophone Taiwan and Beyond by Chia-rong Wu

Cosmopolitanism in China edited by Minghui Hu and Johan Elverskog

The Immortal Maiden Equal to Heaven *and Other Precious Scrolls from Western Gansu* by Wilt L. Idema

Chinese Ethnic Minority Oral Traditions: A Recovered Text of Bai Folk Songs in a Sinoxenic Script by Jingqi Fu and Zhao Min with Xu Lin and Duan Ling

China and Beyond in the Mediaeval Period: Cultural Crossings and Inter-Regional Connections edited by Dorothy C. Wong and Gustav Heldt

Anglophone Literatures in the Asian Diaspora: Literary Transnationalism and Translingual Migrations by Karen An-hwei Lee

Modern Poetry in China: A Visual-Verbal Dynamic by Paul Manfredi

Sinophone Malaysian Literature: Not Made in China by Alison M. Groppe

Infected Korean Language, Purity versus Hybridity: From the Sinographic Cosmopolis to Japanese Colonialism to Global English by Koh Jongsok (translated by Ross King)

The Chinese Prose Poem: A Study of Lu Xun's Wild Grass (Yecao) by Nicholas A. Kaldis

Gao Xingjian: Aesthetics and Creation by Gao Xingjian (translated by Mabel Lee)

Rethinking Chineseness: Translational Sinophone Identities in the Nanyang Literary World by E. K. Tan

A Study of Two Classics: A Cultural Critique of The Romance of the Three Kingdoms and The Water Margin by Liu Zaifu (translated by Shu Yunzhong)

Confucian Prophet: Political Thought in Du Fu's Poetry (752–757) by David K. Schneider

The Classic of Changes in Cultural Context: A Textual Archaeology of the Yi jing by Scott Davis

www.ingramcontent.com/pod-product-compliance
Lightning Source LLC
LaVergne TN
LVHW092116300926
849161LV00002B/152